Critical Infrastructure Security

Cybersecurity lessons learned from real-world breaches

Soledad Antelada Toledano

Critical Infrastructure Security

Group Product Manager: Pavan Ramchandani

Publishing Product Manager: Neha Sharma

Book Project Manager: Ashwin Kharwa

Senior Editors: Arun Nadar, Sayali Pingale

Technical Editor: Irfa Ansari

Copy Editor: Safis Editing

Indexer: Hemangini Bari

Production Designer: Jyoti Kadam

Senior Developer Relations Marketing Executive: Marylou De Mello

First published: May 2024

Production reference: 1300424

Published by Packt Publishing Ltd.

Grosvenor House

11 St Paul's Square

Birmingham

B3 1RB, UK

ISBN 978-1-83763-503-0

www.packtpub.com

To my family, friends, colleagues, and mentors.

Your support, guidance, and belief in me have been invaluable in my journey through the world of cybersecurity. This book, " Critical Infrastructure Security," is a tribute to your unwavering faith and encouragement, which have been instrumental in overcoming challenges and barriers, especially as a woman in this field. I am deeply grateful for your role in making this achievement possible.

– Soledad

Contributors

About the author

Soledad Antelada Toledano, a leading cybersecurity trailblazer, currently serves as security technical program manager at the Office of the CISO at Google. Her career took off at Berkeley Lab, a key player in internet development and scientific research, where she also contributed significantly to NERSC's cybersecurity. Soledad further made her mark as the head of security for the ACM/IEEE Supercomputing Conference, overseeing SCinet's network architecture. She founded GirlsCanHack, advocating for women in cybersecurity. Recognized as one of the 20 Most Influential Latinos in Technology in America in 2016, Soledad is a notable figure in promoting diversity and innovation in cybersecurity.

About the reviewers

Aditya K Sood (Ph.D.) is a cybersecurity leader, advisor, practitioner, and researcher. With the experience of more than 16 years, he provides strategic leadership in the field of information security. Dr. Sood obtained his Ph.D. in computer sciences from Michigan State University. Dr. Sood is also the author of the *Targeted Cyber Attacks* and *Empirical Cloud Security* books. He has been an active speaker and presented at Blackhat, DEFCON, FIRST, APWG, and many others. On the professional front, Dr. Sood held positions such as senior director of threat research and security strategy, director of cloud security, and chief architect while working for companies such as F5 Networks, Symantec, Blue Coat, Elastica, IOActive, and KPMG.

I would like to express my deepest gratitude to all those who contributed to the creation of this book. I am indebted to my family members and mentor for their unwavering support, understanding, and patience throughout the review process. Their encouragement has been a constant source of inspiration.

Chandan Singh Kumbhawat, a cybersecurity maestro with over a decade of experience, specializes in safeguarding critical infrastructure, particularly in the railway sector. He has navigated the complexities of the railway sector, demonstrating a commitment to excellence. His strategic vision and hands-on expertise have fortified systems against evolving threats. Chandan's leadership extends beyond technology, fostering collaboration and knowledge sharing. A trailblazer in adopting cutting-edge tech, he navigates the complex intersection of innovation and cybersecurity, leaving an indelible mark on the industry.

I extend my heartfelt gratitude to my wife and daughters, whose unwavering support and understanding have been the pillars that allowed me to dedicate time and effort to the creation of this book. Their encouragement and sacrifices have been instrumental in shaping this endeavor, and for that, I am truly thankful.

Jean Michel, a seasoned cybersecurity leader with over 2 decades of expertise, specializes in data protection and information security, particularly in critical infrastructure sectors. His strategic roles have driven significant digital transformation and bolstered cyber resilience in urban transport. Renowned for his deep understanding in governance, cyber risk management, and compliance, Jean Michel has been instrumental in safeguarding essential services. His certifications from prestigious bodies underscore his profound knowledge and commitment. As a mentor and innovator, he shapes cybersecurity futures.

Table of Contents

Part 1: Introduction to Critical Infrastructure and Cybersecurity Concepts

1

2

The Growing Threat of Cyberattacks on Critical Infrastructure   31

3

Critical Infrastructure Vulnerabilities   51

Part 2: Dissecting Cyberattacks on CI

4

The Most Common Attacks Against CI   75

5

Analysis of the Top Cyberattacks on Critical Infrastructure   109

Part 3: Protecting Critical Infrastructure

6

Protecting Critical Infrastructure – Part 1   137

7

Protecting Critical Infrastructure – Part 2   167

8

Protecting Critical Infrastructure – Part 3   199

Part 4: What's Next

9

The Future of CI   221

Index   239

Other Books You May Enjoy   248

Preface

This book offers an essential guide for anyone aiming to fortify critical infrastructure against cyber threats. It merges fundamental cybersecurity principles with compelling real-world case studies, enhancing retention and offering engaging insights into the complexities of critical infrastructure cybersecurity. The book specifically addresses the knowledge gap brought about by the convergence of **Information Technology (IT)** and **Operational Technology (OT)**, providing valuable perspective for practitioners navigating this evolving landscape.

It serves as an invaluable resource for cyber defenders, delivering practical knowledge gained from historical cyber incidents to prevent future breaches. From exploring vulnerabilities to presenting strategies for protection, this book equips readers with the understanding necessary to mitigate attacks on critical infrastructure.

You will learn to do the following:

- Comprehend the importance of critical infrastructure and its role within a nation
- Grasp key cybersecurity concepts and terminology
- Recognize the increasing threat of cyberattacks on vital systems
- Identify and understand the vulnerabilities present in critical infrastructure
- Acquire knowledge about the most prevalent cyberattacks targeting these infrastructures
- Implement techniques and strategies to shield critical assets from cyber threats
- Contemplate the future direction of critical infrastructure protection and cybersecurity
- Stay abreast of emerging trends and technologies that may influence security
- Foresee expert predictions on how cyber threats could evolve in the upcoming years
- Gain technical knowledge about the most important cyberattacks in the last years

By the conclusion of this book, you will be well versed in core cybersecurity principles that are instrumental in preventing a broad range of attacks on critical infrastructures.

Who this book is for

This book is designed for a broad audience that includes the following:

- The general public, especially those interested in understanding how cybersecurity issues affect society

- Security enthusiasts who are keen on diving deeper into the specifics of cyber threats and protection measures

- Professionals in the field of cybersecurity or related fields looking for a more nuanced understanding of cyberattacks on critical infrastructure

- Decision-makers and individuals in positions of power with influence over national security policies that want to be informed about the challenges and solutions related to cybersecurity

This book caters to readers with varying levels of pre-existing knowledge, from those with basic understanding to professionals seeking to expand their expertise. It addresses common hurdles for readers, such as unfamiliarity with security concepts, difficulty with technical jargon, and anxiety about the subject matter by breaking down complex ideas into more accessible language and adopting a storytelling approach. The book positions itself uniquely in the market by offering up-to-date insights into the increasing threats of cyberattacks on critical infrastructure, an area where current literature is limited.

What this book covers

Chapter 1, What is Critical Infrastructure?, details the 16 essential CI sectors identified by CISA, such as the chemical and electrical grid sectors, and explains their significance to U.S. national security and safety. It provides an overview of these sectors and examines the potential consequences of cyberattacks, aiming to educate readers on the importance of CI protection and the scenarios of cyber threats.

Chapter 2, The Growing Threat of Cyberattacks on Critical Infrastructure, examines the normalization of cyberattacks on CI, highlighting well-known and obscure cases from recent decades. It investigates the evolution, causes, and emerging trends of these attacks, alongside the intentions behind them, providing a historical context and an evaluation of the current global cybersecurity climate. The chapter aims to enhance the reader's understanding of cybersecurity's development in relation to CI and the landscape of threats from malicious actors on a global scale.

Chapter 3, Critical Infrastructure Vulnerabilities, delves into security vulnerability assessment methods, describing the life cycle of vulnerabilities and the processes for assessing and managing them. It offers insights into prevalent vulnerabilities and threats in critical infrastructure, such as those associated with industrial legacy systems. The chapter clarifies concepts of threats and vulnerabilities, and readers will learn the essentials of vulnerability assessment, how to discern between risk, vulnerability, and threat, becoming familiar with the most common threats and vulnerabilities that affect critical infrastructure today.

Chapter 4, The Most Common Attacks Against CI, offers an in-depth analysis of prevalent cyberattacks targeting critical infrastructure globally. It explores the mechanisms, operations, and success strategies of various attacks such as DDoS, ransomware, supply chain attacks, phishing, unpatched vulnerability exploits, and advanced persistent threats. The chapter is designed to equip readers with detailed technical knowledge of different cyberattacks and an understanding of the attackers' profiles and their objectives.

Chapter 5, Analysis of the Top Cyberattacks on Critical Infrastructure, presents real case studies of cyberattacks aimed at critical sectors. Building upon the foundational knowledge established in the preceding chapters, this chapter offers an in-depth look at the cyberattack landscape, enhancing the reader's technical understanding of such incidents. The focus is on dissecting examples of attacks against national infrastructures and delving into the technical methods employed by attackers. Readers will refine their grasp of cyberattack strategies on CI and learn to apply theoretical insights to real-world scenarios.

Chapter 6, Protecting Critical Infrastructure – Part 1, ventures into the strategies and solutions crucial for safeguarding our essential services from cyber threats. After exposing the potent impact of notable cyber incidents in the previous chapters, this segment turns to proactive defenses. It outlines a range of protective measures, from technical to organizational, vital for reinforcing our critical infrastructure's cybersecurity. The chapter's focus includes network security, continuous monitoring, and the implementation of robust security policies and frameworks.

Chapter 7, Protecting Critical Infrastructure – Part 2, advances the discussion from foundational cybersecurity measures to an in-depth analysis of systems security and endpoint protection. It provides a comprehensive understanding of safeguarding the intricate components of critical infrastructure against advanced cyber threats. The chapter emphasizes robust endpoint security strategies, including the deployment of antivirus and antimalware solutions, and endpoint detection and response systems. It also tackles application security, integrating these security facets into a wider cybersecurity strategy for robust digital protection. This chapter stresses the importance of a layered defense approach in securing critical digital assets amidst the complexity of modern cyber threats.

Chapter 8, Protecting Critical Infrastructure – Part 3, moves beyond proactive measures into the realms of incident response, the cultivation of security culture and awareness, and the role of executive orders in fortifying our critical infrastructure. This part of the series equips the reader with strategies for swift and effective action against security breaches, ensuring infrastructure resilience. Emphasizing the human element, it delves into how fostering a vigilant security-aware culture within organizations contributes to national defense. Additionally, the chapter examines the significant impact of governmental directives on security practices, exploring the intricacies of implementing such orders. This chapter stitches together the practical, cultural, and regulatory facets that are pivotal for the security and readiness of our critical infrastructure.

Chapter 9, The Future of CI, explores the existing shortcomings and the progression in cybersecurity as it pertains to critical infrastructure. It also projects forward to examine the challenges and risks presented by emerging technologies such as artificial intelligence and quantum computing, especially to outdated systems. This chapter contemplates the cybersecurity trajectory and anticipates the resilience needed for critical infrastructures to withstand future threats.

Conventions used

There are a number of text conventions used throughout this book.

Bold: Indicates a new term, an important word, or words that you see onscreen. For example, words in menus or dialog boxes appear in the text like this. Here is an example: "Select **System info** from the **Administration** panel."

> **Tips or important notes**
> Appear like this.

Get in touch

Feedback from our readers is always welcome.

General feedback: If you have questions about any aspect of this book, mention the book title in the subject of your message and email us at customercare@packtpub.com.

Errata: Although we have taken every care to ensure the accuracy of our content, mistakes do happen. If you have found a mistake in this book, we would be grateful if you would report this to us. Please visit www.packtpub.com/support/errata, selecting your book, clicking on the Errata Submission Form link, and entering the details.

Piracy: If you come across any illegal copies of our works in any form on the Internet, we would be grateful if you would provide us with the location address or website name. Please contact us at copyright@packtpub.com with a link to the material.

If you are interested in becoming an author: If there is a topic that you have expertise in and you are interested in either writing or contributing to a book, please visit authors.packtpub.com.

Share Your Thoughts

Once you've read *Critical Infrastructure Security*, we'd love to hear your thoughts! Scan the QR code below to go straight to the Amazon review page for this book and share your feedback.

https://packt.link/r/183763503X

Your review is important to us and the tech community and will help us make sure we're delivering excellent quality content.

Download a free PDF copy of this book

Thanks for purchasing this book!

Do you like to read on the go but are unable to carry your print books everywhere?

Is your eBook purchase not compatible with the device of your choice?

Don't worry, now with every Packt book you get a DRM-free PDF version of that book at no cost.

Read anywhere, any place, on any device. Search, copy, and paste code from your favorite technical books directly into your application.

The perks don't stop there, you can get exclusive access to discounts, newsletters, and great free content in your inbox daily

Follow these simple steps to get the benefits:

1. Scan the QR code or visit the link below

https://packt.link/free-ebook/9781837635030

2. Submit your proof of purchase

3. That's it! We'll send your free PDF and other benefits to your email directly

Part 1:
Introduction to Critical Infrastructure and Cybersecurity Concepts

Part 1 serves as a primer on the fundamental aspects of critical infrastructure and the cyber threats that jeopardize its integrity. It begins with an exploration of the key sectors vital to national security and public safety, discussing the potential impact of cyber incidents. The discussion then shifts to the evolution of cyber threats, offering insights into the historical context and current trends that shape the cybersecurity landscape. Lastly, it addresses the methodologies for identifying and mitigating vulnerabilities, with a special focus on the unique challenges faced by industrial legacy systems. This section establishes the groundwork for understanding the complex world of cybersecurity and the strategies needed to protect critical infrastructure.

This part has the following chapters:

- *Chapter 1, What is Critical Infrastructure?*
- *Chapter 2, The Growing Threat of Cyberattacks on Critical Infrastructure*
- *Chapter 3, Critical Infrastructure Vulnerabilities*

What is Critical Infrastructure?

Critical infrastructure (CI) refers to the assets, systems, and networks that are essential for the functioning of a society and its economy. These include physical assets that support the delivery of services such as energy, water, transportation, healthcare, communications, emergency services, and financial services. The term critical infrastructure also encompasses the resources, facilities, and systems that are necessary for national security, public safety, and public health.

The **Cybersecurity and Infrastructure Security Agency** (**CISA**) identifies 16 CI sectors in the United States, as shown in *Figure 1.1*. These sectors are considered so vital that their disruption, incapacitation, or destruction could have a severe impact on national security, public health and safety, or economic security:

Figure 1.1 – Critical infrastructure sector

This chapter will cover the following topics:

- Overview of CI sectors
- Impacts of compromised sectors
- Cyberattack scenarios in CI sectors
- Risk mitigation examples

To shift our focus toward a more detailed examination of each sector, let's now explore them individually.

Chemical sector

The **chemical sector** is one of the 16 CI sectors identified by the CISA in the United States. It includes the production, storage, and transportation of chemicals that are essential to many industries, such as agriculture, healthcare, and manufacturing. The sector is diverse, including companies that produce industrial chemicals, pesticides, pharmaceuticals, and other specialty chemicals. The chemical sector is vital to the U.S. economy, and a disturbance in its functioning could lead to serious implications for public health, safety, and the security of the nation.

Impact of a compromised chemical sector

If the chemical sector were compromised or under attack, it could have severe consequences. For example, a cyberattack on a chemical plant could result in the release of toxic chemicals into the environment, causing harm to people, animals, and plants. A disruption to the production of chemicals could also impact other CI sectors, such as the healthcare sector, which relies on pharmaceuticals and medical devices. Additionally, the chemical sector plays a critical role in the supply chain for many industries, and a disruption to its operations could have ripple effects throughout the economy.

Cyberattack scenarios in the chemical sector

The chemical sector, vital for manufacturing and supplying essential chemicals, faces critical cyberattack scenarios that can result in operational disruptions, environmental hazards, and national security risks. Here are some key cyberattack scenarios that necessitate heightened security measures and proactive defense strategies in this sector:

- **Ransomware attack**: A ransomware attack could target a chemical plant's control systems, which could cause the plant to shut down or release toxic chemicals into the environment. The attackers could then demand a ransom payment in exchange for the safe return of control of the systems.

- **Supply chain attack**: A cyberattack on a chemical supplier could impact the production of essential chemicals, which could have a ripple effect throughout the economy. Attackers could target the supplier's systems to steal intellectual property or disrupt operations, leading to shortages of critical chemicals.

- **Insider threat**: A malicious insider could use their access to a chemical plant's control systems to cause damage or release toxic chemicals. This could be done for financial gain or to cause harm to the company or its employees.

- **State-sponsored cyberattack**: A nation-state could target the chemical sector to disrupt the production of critical chemicals or to steal intellectual property for use in their industries. Such an attack could have severe consequences on national security and economic stability.

- **Internet of Things (IoT) attack:** IoT devices are increasingly used in the chemical sector to monitor production processes and control systems. A cyberattack on these devices could compromise the entire system, leading to a shutdown or release of toxic chemicals. Attackers could use the compromised devices to launch further attacks or to steal sensitive data.

The chemical sector is an essential component of the U.S. economy, and its operations are critical to many other sectors. A disruption to its operations due to a cyberattack could have severe consequences on public health, safety, and national security. Therefore, it is essential to protect and secure the chemical sector's assets, systems, and networks against cyber threats.

Commercial facilities sector

The **commercial facilities sector** is another one of the 16 CI sectors identified by the CISA in the United States. This sector includes a wide range of facilities, such as office buildings, shopping malls, sports stadiums, and entertainment venues. It also includes facilities that provide essential services, such as transportation hubs, hotels, and restaurants. The sector is essential to the functioning of society, and a disruption to its operations could have severe consequences on public safety and economic stability.

Impact of a compromised commercial facilities sector

If the commercial facilities sector were compromised or under attack, it could have severe consequences:

- **Economic disruption**: A cyberattack on transportation hubs or commercial facilities can disrupt the flow of goods and people, resulting in significant economic losses. It can hamper business operations, affect supply chains, and lead to financial repercussions for businesses and the broader economy.

- **Public safety concerns**: Attacks on sports stadiums or entertainment venues can jeopardize public safety, potentially leading to the cancellation or disruption of events. This can have a negative impact on attendees and the reputation of the facility, causing a loss of trust among the public.

- **Data breaches and financial loss**: Cyberattacks targeting hotel or restaurant chains can compromise sensitive data, including credit card information and personal details of customers. Such breaches can lead to financial loss due to fraud, legal liabilities, and damage to the brands' reputation. Restoring trust and recovering from a data breach can be time-consuming and costly.

- **Reputational damage**: A compromised commercial facilities sector can result in significant reputational damage for businesses. News of cyberattacks or data breaches can erode customer trust, leading to a decline in patronage and potential long-term consequences for the affected companies' brand image.

- **Legal and regulatory implications**: A cyberattack on commercial facilities may result in legal and regulatory consequences. Depending on the jurisdiction, businesses may be subject to fines, penalties, or legal action for failing to adequately protect customer data or maintain adequate cybersecurity measures.

To mitigate these risks, it is crucial for commercial facilities to implement robust cybersecurity measures, regularly update systems, conduct employee training, and have effective incident response plans in place.

Cyberattack scenarios in the commercial facilities sector

The commercial facilities sector, comprising various establishments such as hotels, restaurants, transportation hubs, and sports stadiums, is vulnerable to cyberattacks that can disrupt operations, compromise sensitive data, and undermine customer trust. Here are some critical cyberattack scenarios that pose significant risks to this sector:

- **Ransomware attack**: A ransomware attack could target a chain of hotels or restaurants, which could result in the theft of sensitive data and the encryption of critical systems. The attackers could then demand a ransom payment in exchange for the safe return of control of the systems and the data.

- **Insider threat**: A malicious insider could use their access to a commercial facility's systems to cause damage or steal sensitive data. This could be done for financial gain or to cause harm to the company or its customers.

- **Distributed denial of service (DDoS) attack**: A DDoS attack could target a transportation hub's or sports stadium's website, causing it to crash and preventing people from accessing critical information. The attack could also disrupt the facility's operations by overwhelming its network with traffic.

- **Social engineering attack**: A social engineering attack could target employees of a commercial facility, tricking them into divulging sensitive information or granting access to critical systems. The attackers could then use this information to launch further attacks or steal sensitive data.

- **Internet of Things (IoT) attack**: IoT devices are increasingly used in commercial facilities to monitor operations and provide services to customers. A cyberattack on these devices could compromise the entire system, leading to a shutdown of operations or a breach of sensitive data. Attackers could use the compromised devices to launch further attacks or to steal sensitive data.

Ensuring robust cybersecurity measures and comprehensive employee training is essential for the commercial facilities sector to mitigate the risks of ransomware attacks, insider threats, DDoS attacks, social engineering, and IoT vulnerabilities, safeguarding operations, data, and customer trust.

Communications sector

The **communications sector** refers to the systems and networks that enable the transmission of information, including voice, data, and video, across various platforms. This sector includes wired and wireless communication networks, broadcasting systems, satellite systems, and internet service providers. The communications sector is essential for the functioning of many other CI sectors, including the energy, transportation, and financial sectors, and any disruption in this sector can have far-reaching consequences.

Impact of a compromised communications sector

If the communications sector were compromised or under attack, there would be significant disruptions to the functioning of many other CI sectors. For example, emergency responders rely on communication networks to coordinate their response efforts, and any disruption to these networks could impede their ability to effectively respond to emergencies. Disruptions to communication networks could also lead to disruptions in the supply chain, as logistics companies rely on these networks to track shipments and coordinate deliveries.

Cyberattack scenarios in the communications sector

There are several potential cyberattack scenarios that could target the communications sector. One such scenario is a DDoS attack, in which a network of compromised devices, known as a botnet, floods communication networks with traffic, making them inaccessible to legitimate users. Another scenario is a **person-in-the-middle attack**, in which an attacker intercepts communications between two parties and can either eavesdrop on the communication or modify it for their own purposes. A third scenario is a ransomware attack, in which an attacker encrypts critical data and demands payment in exchange for the decryption key. These are just a few examples of the many potential cyberattack scenarios that could target the communications sector. It is essential for organizations in this sector to take appropriate cybersecurity measures to prevent and mitigate the impact of these attacks.

Critical manufacturing sector

The **critical manufacturing sector** encompasses industries involved in producing essential goods and materials such as automobiles, aerospace products, electronics, pharmaceuticals, and chemicals. It plays a vital role in the economy, national security, and public well-being by ensuring the availability of essential products. This sector relies heavily on advanced technologies, automation, and interconnected systems to optimize production processes and supply chains.

Impact of a compromised critical manufacturing sector

If the critical manufacturing sector were compromised or under attack, it could have severe consequences on various levels:

Economic disruption	Disruptions in critical manufacturing operations can lead to supply chain disruptions, product shortages, and increased costs, affecting both businesses and consumers. This can have a cascading effect on the overall economy.
National security threats	Compromised critical manufacturing facilities may result in the loss of sensitive intellectual property, jeopardizing national security interests. Additionally, essential defense-related products and equipment may become unavailable, affecting military readiness.
Public safety concerns	Attacks on critical manufacturing systems can impact the safety and quality of products. Malicious actors may manipulate production processes, leading to defective or unsafe goods that could pose risks to public health and safety

Table 1.1 – Implications of a compromised critical manufacturing sector

A compromise of the critical manufacturing sector poses significant risks, including economic disruption, national security threats, and public safety concerns, emphasizing the importance of safeguarding this sector against cyberattacks.

Cyberattack scenarios in the critical manufacturing sector

The critical manufacturing sector is vulnerable to various cyberattack scenarios that can disrupt operations, compromise intellectual property, and exploit insider threats. Here are some key scenarios to be aware of:

- **Ransomware attack**: A cybercriminal could deploy ransomware to disrupt critical manufacturing operations by encrypting data and systems, demanding a ransom to restore access. This could halt production, disrupt supply chains, and result in financial losses.

- **Supply chain attack**: Adversaries may target suppliers or subcontractors within the critical manufacturing sector, exploiting vulnerabilities in their systems to gain unauthorized access. This can provide attackers with a pathway to infiltrate and compromise larger manufacturing networks.

- **Intellectual property theft**: Nation-state actors or competitors may launch sophisticated cyber espionage campaigns to steal proprietary manufacturing processes, designs, or trade secrets. This could result in significant economic losses and undermine the competitiveness of the affected companies.

- **Insider threats**: Insider threats pose a risk within the critical manufacturing sector. Disgruntled employees or insiders with authorized access could sabotage production systems, compromise sensitive information, or leak valuable intellectual property.

To mitigate the risks and consequences of cyberattacks on the critical manufacturing sector, it is crucial for companies to implement robust cybersecurity measures, such as network segmentation, regular system patching, employee training on phishing and social engineering, and continuous monitoring of IT systems. Collaboration between government agencies, industry stakeholders, and cybersecurity experts is also essential in developing and implementing effective strategies to protect critical manufacturing infrastructure.

Dams sector

The **dams sector** refers to the infrastructure and systems involved in the construction, operation, and maintenance of dams and associated facilities. Dams play a crucial role in water resource management, hydroelectric power generation, flood control, and irrigation. They provide a reliable water supply and contribute to the economic and social development of regions around the world.

Impact of a compromised dams sector

If the dams sector were compromised or under attack, it could have significant consequences on various levels:

- **Infrastructure damage**: Attacks targeting dams could result in physical damage to the structures, such as breaching or destabilizing the dams. This could lead to catastrophic flooding, loss of life, and extensive property damage downstream.

- **Water supply disruptions**: Compromised dams can disrupt water supply systems, affecting drinking water availability, irrigation for agriculture, and industrial water usage. This can have far-reaching consequences for communities, agricultural production, and industrial operations.

- **Power generation disruptions**: Many dams are also associated with hydroelectric power generation. Attacks on dam infrastructure could disrupt power generation, leading to electricity shortages and impacting the stability of regional power grids.

- **Environmental impact**: Dam breaches caused by cyberattacks could release large volumes of water into natural ecosystems, causing significant environmental damage, loss of biodiversity, and disruption to aquatic habitats.

The protection and resilience of the dams sector are crucial to mitigate the potential impacts of a compromised infrastructure. By ensuring robust security measures, regular maintenance, and effective response plans, stakeholders can minimize the risks of infrastructure damage, water supply disruptions, power generation interruptions, and adverse environmental consequences.

Cyberattack scenarios in the dams sector

The dams sector faces various cyberattack scenarios that can pose significant risks to the safety and operational integrity of dams.

- **Remote access exploitation**: Adversaries may attempt to exploit vulnerabilities in the control systems of dams, gaining unauthorized remote access. This can allow attackers to manipulate water release mechanisms, modify operational parameters, or disrupt communication networks.

- **Data manipulation**: Cybercriminals could target the data management systems of dams, altering operational data such as water level measurements or flow rates. This can lead to incorrect decisions being made regarding dam operations, potentially compromising safety and water management.

- **DDoS attacks**: Dams often rely on computer-based systems to manage operations. DDoS attacks can overwhelm these systems with a flood of traffic, causing disruptions in monitoring, control, and communication capabilities.

- **Insider threats**: Insider threats within the dams sector pose a significant risk. Disgruntled employees with authorized access to critical systems could intentionally sabotage or manipulate dam operations, compromising safety and integrity.

Adversaries may exploit vulnerabilities in control systems, manipulate data management systems, launch DDoS attacks, or exploit insider threats. Safeguarding the dams sector against these cyber threats is essential to ensure the reliable and secure operation of dams, protecting public safety and water management.

To mitigate the risks associated with cyberattacks on the dams sector, it is crucial to implement robust cybersecurity measures. This includes regular security assessments, network monitoring, access controls, encryption of sensitive data, employee training on cybersecurity best practices, and close collaboration between dam operators, government agencies, and cybersecurity practicioners. Proactive measures can help identify vulnerabilities, strengthen defenses, and ensure the reliable and secure operation of dams for the benefit of society and the environment.

Defense industrial base sector

The **defense industrial base (DIB) sector** plays a vital role in supporting national defense and military capabilities. Comprised of organizations, contractors, manufacturers, and suppliers involved in the research, development, production, and maintenance of defense-related goods and services, the DIB sector is critical for ensuring the readiness and effectiveness of a nation's defense infrastructure.

Impact of a compromised defense industrial base sector

If the DIB sector were compromised or under attack, the ramifications would be significant. The consequences could range from national security risks to operational disruptions and economic impacts.

One of the primary concerns of a compromised DIB sector is the potential compromise of national security. Adversaries gaining access to sensitive military technologies, classified information, and intellectual property can significantly undermine a nation's defense capabilities. The theft of critical defense technologies and military secrets poses a severe threat to a country's national security and can compromise its military superiority and readiness.

Attacks on the DIB sector can disrupt the production, supply chain, and maintenance of defense systems. Delays in the delivery of equipment and reduced operational readiness can hinder a country's ability to effectively respond to threats and maintain a strong defense posture.

The economic impact of a compromised DIB sector cannot be overlooked. The sector generates jobs, drives innovation, and contributes to the broader industrial base. A compromised DIB sector can result in economic losses, job cuts, and disruptions in the supply chain. The ripple effects can extend beyond defense contractors, affecting the overall economy and stability of industries connected to the DIB sector.

Cyberattack scenarios in the defense industrial base sector

In terms of cyberattack scenarios, several possibilities exist for targeting the DIB sector. **Advanced persistent threats** (**APTs**) are sophisticated, long-term infiltration campaigns orchestrated by state-sponsored attackers. These attacks involve persistent access to sensitive networks, data exfiltration, and the theft of intellectual property, military secrets, and critical defense technologies.

Supply chain attacks pose another significant threat. Adversaries can exploit vulnerabilities in the supply chain by targeting subcontractors, suppliers, or manufacturers within the DIB sector. By compromising these entities, attackers can inject malicious code into defense systems or compromise the integrity of components, resulting in compromised security and functionality.

Insider threats are also a concern. Malicious insiders or unintentional actions by employees with access to sensitive information can lead to the theft of classified data, sabotage of defense systems, or unauthorized disclosure of critical information to adversaries.

Ransomware attacks, where cybercriminals encrypt critical systems and demand ransom for their release, can also impact the DIB sector. Such attacks can disrupt operations, compromise sensitive data, and cause financial losses.

To mitigate these risks, the DIB sector must prioritize robust cybersecurity measures. This includes implementing strong network security protocols, conducting regular security assessments, fostering a culture of cybersecurity awareness, establishing information-sharing partnerships, and investing in advanced threat detection and response capabilities. By doing so, the DIB sector can mitigate risks, safeguard national security, and ensure the continuity of defense operations in the face of evolving cyber threats.

Emergency services sector

The **emergency services sector** is a critical component of any society, encompassing organizations and agencies responsible for responding to and managing emergencies, including law enforcement, fire services, emergency medical services, and disaster response teams. The sector plays a crucial role in safeguarding public safety and well-being during crisis situations. However, if the emergency services sector were compromised or under attack, the consequences would be severe and far-reaching.

Impact of a compromised emergency services sector

One of the primary consequences of a compromised emergency services sector is the potential breakdown of emergency response capabilities. In a cyberattack scenario, vital communication systems could be disrupted, preventing effective coordination between emergency personnel and agencies. This disruption can hinder the ability to respond promptly and efficiently to emergencies, resulting in delays in critical assistance and potentially escalating the severity of the situation.

Another significant concern is the potential compromise of sensitive information and systems. Emergency services hold a vast amount of personal data, including medical records, contact details, and confidential information related to ongoing investigations. If these systems are compromised, it can lead to the exposure of sensitive information, violating privacy rights and potentially endangering individuals involved in emergency situations.

Cyberattack scenarios in the emergency services sector

Cyberattack scenarios targeting the emergency services sector can take various forms. One such scenario involves DDoS attacks, where attackers overload communication systems with a flood of traffic, rendering them unavailable. In such instances, emergency personnel would struggle to access critical information and communicate effectively, significantly hampering their response capabilities.

Ransomware attacks pose another significant threat to the emergency services sector. Attackers can infiltrate systems and encrypt vital data and systems, demanding a ransom for their release. If successful, these attacks can disrupt operations, paralyze emergency response efforts, and potentially compromise sensitive data.

Phishing attacks also pose a risk to the sector. Attackers can impersonate trusted individuals or organizations and attempt to deceive emergency personnel into revealing sensitive information or providing unauthorized access to systems. Successful phishing attacks can result in unauthorized access to CI, compromise of communication channels, or the deployment of malicious software.

To mitigate the risks and consequences of cyberattacks on the emergency services sector, robust cybersecurity measures must be in place. This includes implementing advanced firewalls, intrusion detection systems, and encryption protocols to protect sensitive data and communication channels. Regular training and awareness programs should be conducted to educate personnel about potential cyber threats and best practices for safeguarding information. Collaboration with cybersecurity experts and information sharing among agencies can help identify and respond to emerging threats effectively.

In conclusion, the emergency services sector is a vital component of public safety and requires strong cybersecurity measures to protect its critical systems and information. The consequences of a compromised emergency services sector can lead to delays in emergency response, exposure of sensitive data, and potential harm to individuals. By investing in cybersecurity and adopting proactive measures, the emergency services sector can enhance its resilience and continue to fulfill its crucial role in safeguarding communities during times of crisis.

Energy sector

The **energy sector** plays a critical role in powering economies, providing electricity, and fueling transportation. It encompasses various subsectors, including oil and gas, electric power generation, renewable energy, and nuclear power. As our reliance on technology and interconnected systems increases, the energy sector faces growing cybersecurity challenges and potential threats. A compromise or attack on this sector can have severe consequences, affecting not only the industry but also the economy and public safety.

Impact of a compromised energy sector

A compromise of the energy sector can have profound impacts on energy supply, economies, and infrastructure. Here are some key consequences that can arise from a compromised energy sector:

- **Disruption in energy supply**: If the energy sector is compromised or under attack, it can lead to disruptions in energy supply. Power outages, shutdowns of oil and gas refineries, or disruption of renewable energy generation can result in significant economic losses, inconvenience to businesses and individuals, and potential risks to public safety.

- **Economic consequences**: The energy sector is a vital component of economic stability and growth. An attack that disrupts energy production, distribution, or pricing mechanisms can have far-reaching economic consequences, including increased costs for businesses and consumers, loss of revenue, and decreased productivity.

- **Infrastructure damage**: Cyberattacks targeting energy infrastructure can cause physical damage to critical systems and equipment. For example, an attack on a power grid could damage transformers or control systems, leading to extended downtime, costly repairs, and potential safety hazards.

A compromise of the energy sector can have devastating effects on energy supply, economies, and infrastructure. Disruptions in energy supply can result in significant economic losses, inconvenience to businesses and individuals, and potential risks to public safety. Moreover, the economic consequences of an attack on energy production, distribution, or pricing mechanisms can lead to increased costs, loss of revenue, and decreased productivity. Cyberattacks targeting energy infrastructure can cause physical damage, such as damage to transformers or control systems, resulting in extended downtime, costly repairs, and potential safety hazards.

Cyberattack scenarios in the energy sector

The energy sector faces a growing threat from cyberattacks, with various attack scenarios capable of causing severe disruptions, compromising sensitive information, and jeopardizing operations. Here are some key cyberattack scenarios that pose significant risks to the energy sector:

- **Ransomware attacks**: In a ransomware attack, malicious actors can infiltrate energy companies' networks and encrypt critical files and systems. They then demand a ransom in exchange for restoring access. Such attacks can paralyze operations, disrupt energy supply, and result in significant financial losses.

- **APTs**: APTs involve sophisticated and prolonged attacks by well-funded and organized adversaries. In the energy sector, APTs may target sensitive information, intellectual property, or control systems to gain unauthorized access, gather intelligence, or sabotage operations.

- **Insider threats**: Insider threats pose a significant risk in the energy sector, as malicious insiders or employees with compromised credentials can exploit their privileged access to compromise critical systems, steal sensitive data, or cause intentional damage.

- **DDoS attacks**: DDoS attacks overwhelm energy company networks or websites by flooding them with an excessive volume of traffic. This can disrupt online services, hinder communications, and impact customer access to energy-related services.

In conclusion, the energy sector faces a multifaceted and evolving threat landscape in terms of cyberattacks. Ransomware attacks, APTs, insider threats, and DDoS attacks pose substantial risks to the sector's operations, infrastructure, and the security of sensitive information.

Preventing and mitigating cyberattacks

To enhance the security posture of the energy sector, several of the following measures can be implemented:

- **Strong cybersecurity practices**: Energy companies should adopt robust cybersecurity practices, including regular vulnerability assessments, network monitoring, and incident response planning. It is crucial to keep systems and software up to date with the latest patches and security updates.

- **Employee education and training**: Training programs should be conducted to educate employees about cybersecurity best practices, such as recognizing phishing emails, using strong passwords, and safeguarding sensitive information.

- **Enhanced network segmentation**: Implementing proper network segmentation isolates critical systems, reducing the potential for lateral movement by attackers and limiting the impact of a compromise.

- **Continuous monitoring and threat intelligence**: The energy sector should utilize advanced monitoring tools and threat intelligence to identify and respond to cyber threats in real time. Intrusion detection systems, **security information and event management (SIEM)** systems, and threat intelligence feeds can provide valuable insights.

- **Collaboration and information sharing**: The energy sector should foster collaboration among industry stakeholders, government agencies, and cybersecurity organizations to share threat intelligence and best practices, and collaborate on incident response.

Enhancing the security of the energy sector against cyberattacks requires a multi-faceted approach, including robust cybersecurity practices, employee education, network segmentation, continuous monitoring, and collaborative information sharing among stakeholders. By implementing these measures, the energy sector can better prevent and mitigate cyber threats, safeguarding CI and ensuring the reliability and resilience of energy systems.

Financial services sector

The **financial services sector** plays a crucial role in the global economy, encompassing a wide range of institutions and activities related to financial transactions, investments, and monetary management. It includes banks, insurance companies, asset management firms, stock exchanges, and other financial intermediaries. The sector facilitates the flow of capital, provides essential services to individuals and businesses, and contributes to economic growth and stability.

Impact of a compromised financial services sector

If the financial services sector were compromised or under attack, significant consequences could occur on both a national and global scale. Some potential impacts include the following:

- **Economic disruption**: A compromise or attack on the financial services sector can disrupt the functioning of financial markets, leading to volatility, reduced investor confidence, and potential economic downturns. It can affect stock prices, currency exchange rates, interest rates, and the availability of credit, impacting businesses and individuals alike.

- **Financial losses**: Attacks targeting financial institutions can result in financial losses due to theft, fraud, or unauthorized access to sensitive information. These losses can occur at both institutional and individual levels, potentially affecting savings, investments, and financial stability.

- **Customer trust and reputation**: A compromised financial services sector can erode customer trust and confidence in the security of financial systems. Customers may hesitate to conduct transactions or share sensitive information, impacting the overall functioning of the sector. Financial institutions may also face reputational damage, which can have long-term consequences on their business operations.

- **Regulatory compliance**: Cyberattacks on the financial services sector can lead to regulatory compliance breaches, violating data protection and privacy regulations. Institutions may face legal consequences, fines, and penalties for failing to adequately protect customer information or comply with industry standards.

Cyberattack scenarios in the financial services sector

Several cyberattack scenarios that pose risks to the financial services sector include the following:

- **DDoS attacks**: Attackers can target financial institutions' websites and systems with massive traffic to overwhelm their servers, causing service disruptions and rendering online banking and financial services inaccessible to customers.

- **Phishing and social engineering**: Cybercriminals can send fraudulent emails or messages, posing as legitimate financial institutions, to deceive customers into sharing sensitive information such as login credentials or personal details. This information can then be used for unauthorized access or identity theft.

- **Insider threats**: Malicious insiders with access to financial systems and customer data can exploit their privileges to steal sensitive information, manipulate transactions, or disrupt operations. This can include employees, contractors, or third-party vendors with authorized access.

- **APTs**: Sophisticated and persistent cyberattacks targeting financial institutions involve long-term infiltration, stealthy data exfiltration, and targeted attacks to compromise critical systems. APTs can be orchestrated by state-sponsored actors, organized crime groups, or highly skilled malicious hackers.

- **Ransomware attacks**: Financial institutions can be targeted by ransomware, where attackers encrypt critical data and demand a ransom for its release. This can lead to data loss, operational disruptions, and financial losses if institutions are unable to recover the encrypted data or pay the ransom.

To mitigate the risks and protect the financial services sector from cyberattacks, institutions should implement robust cybersecurity measures, including network security, encryption, access controls, threat intelligence, employee training, incident response plans, and regular security assessments. Collaboration between financial institutions, regulatory bodies, and law enforcement agencies is also crucial to ensure effective defense against cyber threats and to maintain the stability and security of the financial services sector.

Food and agriculture services sector

The **food and agriculture services sector** encompasses a wide range of activities related to the production, processing, distribution, and retailing of food and agricultural products. It includes agricultural farms, food processing plants, wholesalers, retailers, and various service providers supporting the sector. This sector plays a vital role in ensuring food security, supporting rural livelihoods, and meeting the nutritional needs of the population.

Impact of a compromised food and agriculture sector

If the food and agriculture services sector were compromised or under attack, significant consequences could occur, affecting both the economy and public health. Some potential impacts include the following:

- **Disruption in food supply chains**: Cyberattacks on the food and agriculture sector can disrupt the entire supply chain, leading to shortages, price fluctuations, and compromised food safety. Attackers may target CI, logistics systems, or information systems, hindering the movement of goods and causing delays in production and distribution.

- **Compromised food safety**: An attack on the sector's information systems can lead to the manipulation or alteration of food safety data, making it difficult to identify and mitigate potential risks. This can result in the distribution and consumption of contaminated or unsafe food, posing risks to public health and potentially leading to foodborne illnesses.

- **Financial losses**: Attacks targeting financial transactions and systems within the food and agriculture services sector can result in financial losses for businesses. This can include theft of funds, fraudulent transactions, or disruption of financial operations, impacting the profitability and sustainability of farms, processors, and other businesses within the sector.

- **Damage to reputation**: A compromised food and agriculture services sector can lead to a loss of consumer trust and confidence in the safety and quality of food products. Incidents of contamination, adulteration, or other malicious activities can tarnish the reputation of companies and negatively impact their brand image, resulting in long-term consequences for their business operations.

- **Economic impact**: The food and agriculture services sector is a significant contributor to the economy, both in terms of employment and revenue generation. Compromising this sector can have broader economic implications, affecting rural livelihoods, export opportunities, and overall economic growth.

In conclusion, a compromise of the food and agriculture services sector can have detrimental effects on the economy and public health, including disruptions in food supply chains, compromised food safety, financial losses, damage to reputation, and broader economic impacts.

Cyberattack scenarios in the food and agriculture services sector

Several cyberattack scenarios that pose risks to the food and agriculture services sector include the following:

- **Supply chain disruptions**: Attackers may target the sector's supply chain systems, including inventory management, transportation, and logistics platforms. By disrupting these systems, they can cause delays in product delivery, create shortages, or introduce counterfeit products into the market.

- **Data breaches**: Cybercriminals may attempt to breach the information systems of food and agriculture companies to gain access to sensitive data. This can include customer information, financial records, or proprietary information, which can be used for financial gain or sold on the dark web.

- **Industrial espionage**: Competitors or foreign entities may engage in cyber espionage to steal intellectual property, such as proprietary technologies, research data, or innovative farming techniques. This can undermine the competitive advantage of companies and hinder innovation within the sector.

- **Disruption of CI**: The sector relies on various CIs, such as irrigation systems, storage facilities, and processing plants. Targeting these systems with cyberattacks can disrupt operations, leading to production delays, equipment failures, or even physical damage.

- **Misinformation and social engineering**: Attackers may spread misinformation or engage in social engineering tactics, such as spreading false food safety alerts, manipulating online reviews, or deceiving consumers about the origin or quality of food products. This can create panic, erode consumer trust, and damage the reputations of businesses within the sector.

To mitigate the risks and protect the food and agriculture services sector from cyberattacks, companies should implement robust cybersecurity measures, including secure network infrastructure, regular system updates and patches, employee training on cybersecurity best practices, and incident response plans.

Government facilities sector

The **government facilities sector** encompasses a wide range of services provided by government agencies to support the functioning of public facilities and infrastructure. It includes services such as maintenance, security, transportation, and administrative support for government buildings, public spaces, and CI. This sector plays a crucial role in ensuring the smooth operation of government operations, public services, and the overall functioning of society.

Impact of a compromised government facilities sector

If the government facilities sector were compromised or under attack, significant consequences could occur, affecting both government operations and public safety. Some potential impacts include the following:

- **Disruption of essential services**: Cyberattacks on government facilities can disrupt essential services provided to the public, such as transportation systems, utilities, emergency response services, and administrative functions. This can lead to service interruptions, delays, and decreased efficiency in delivering public services, impacting the daily lives of citizens.

- **Compromised infrastructure**: Attacks targeting government facilities can compromise CI, including power plants, water treatment facilities, transportation hubs, and communication networks. Such attacks can disrupt essential services, lead to infrastructure failures, or even pose risks to public safety.

- **Data breaches and privacy concerns**: Government facilities store a vast amount of sensitive data, including personal information of citizens, classified government documents, and CI blueprints. A cyberattack can result in data breaches, leading to unauthorized access, theft, or exposure of sensitive information. This can have severe implications for national security, privacy, and public trust in the government.

- **Political and economic impact**: A compromised government facilities sector can have significant political and economic consequences. It can undermine public confidence in the government's ability to protect CI and provide essential services. Additionally, the cost of recovering from cyberattacks and implementing stronger security measures can strain government budgets and resources.

In conclusion, a compromise of the government facilities sector can have wide-ranging impacts, including disruptions to essential services, compromised infrastructure, data breaches, and privacy concerns, as well as political and economic ramifications.

Cyberattack scenarios in the government facilities sector

Several cyberattack scenarios that pose risks to the government facilities sector include the following:

- **Ransomware attacks**: Attackers may deploy ransomware on government systems, encrypting critical data and demanding ransom for its release. This can paralyze government operations, disrupt essential services, and force the government to make difficult decisions regarding payment.

- **APTs**: APT groups may target government facilities to gain persistent access to networks and systems. They can infiltrate networks, gather sensitive information, and remain undetected for long periods, potentially compromising CI or conducting espionage activities.

- **Physical infrastructure attacks**: Cyberattacks targeting government facilities may aim to manipulate or disable physical infrastructure systems, such as access control systems, surveillance cameras, or building automation systems. This can compromise security measures, compromise safety protocols, or facilitate unauthorized access to sensitive areas.

- **Social engineering and spear phishing**: Attackers may employ social engineering techniques, such as spear phishing, to deceive government employees into revealing sensitive information or granting unauthorized access to systems. This can lead to unauthorized access to government networks, data breaches, or the spread of malware.

- **Insider threats**: The government facilities sector may face risks from insider threats, where individuals with authorized access to systems intentionally or unintentionally compromise security. This can include unauthorized disclosure of sensitive information, sabotage of systems, or insider attacks aimed at disrupting operations.

To mitigate the risks and protect the government facilities sector from cyberattacks, robust cybersecurity measures are essential. These can include implementing strong access controls, conducting regular security assessments, training employees on cybersecurity best practices, implementing incident response plans, and collaborating with cybersecurity agencies to share threat intelligence and best practices.

Healthcare and public health sector

The **healthcare and public health sector** plays a vital role in providing medical care, public health services, and emergency response to safeguard the well-being of individuals and communities. It encompasses various entities, including hospitals, clinics, medical research facilities, public health agencies, and pharmaceutical companies. This sector is responsible for ensuring the delivery of essential healthcare services, promoting public health, and responding to medical emergencies and outbreaks.

Impact of a compromised healthcare and public health sector

If the healthcare and public health sector were compromised or under attack, it could have severe consequences impacting both individuals and society. Some potential impacts include the following:

- **Disruption of healthcare services**: Cyberattacks on healthcare systems can disrupt critical healthcare services, including patient care, diagnostics, treatment, and medical records management. This can lead to delayed or compromised medical treatments, jeopardizing patient safety and potentially resulting in adverse health outcomes.

- **Compromised patient data and privacy**: Healthcare organizations store vast amounts of sensitive patient data, including medical records, personal information, and billing details. A cyberattack can result in data breaches, exposing confidential patient information to unauthorized access, identity theft, or misuse. Such breaches erode patient trust in the healthcare system and can have legal and financial implications for healthcare providers.

- **Impaired emergency response:** The healthcare sector plays a crucial role in emergency response during public health crises, natural disasters, or disease outbreaks. If compromised, the ability to effectively respond to emergencies, coordinate resources, and provide timely medical care may be severely impacted, leading to increased morbidity and mortality rates.

- **Medical device compromise:** The healthcare sector relies on various medical devices and equipment for patient care and treatment. Cyberattacks can target these devices, compromising their functionality or manipulating their operation. This can result in the delivery of incorrect treatment, device malfunction, or disruption of critical life-supporting systems.

- **Intellectual property theft:** Medical research institutions and pharmaceutical companies are prime targets for cyber espionage and intellectual property theft. Attackers may aim to steal valuable research data, clinical trial information, or proprietary knowledge, leading to financial losses, setbacks in medical advancements, and potential harm to public health.

In conclusion, a compromise of the healthcare and public health sector poses significant risks to patient care, data privacy, emergency response capabilities, medical device functionality, and intellectual property protection.

Cyberattack scenarios in the healthcare and public health sector

Several cyberattack scenarios that pose risks to the healthcare and public health sector include the following:

- **Ransomware attacks:** Cybercriminals may deploy ransomware to encrypt healthcare systems and demand ransom for data decryption. This can paralyze healthcare operations, hinder access to patient records, and delay critical medical procedures, potentially compromising patient safety and care.

- **Data breaches and patient information theft:** Hackers may infiltrate healthcare databases to steal patient information, including medical records, insurance details, and personally identifiable information. This stolen data can be sold on the black market or used for various malicious purposes, leading to identity theft, fraud, or targeted phishing attacks.

- **DDoS attacks:** Attackers may launch DDoS attacks against healthcare websites or systems, overwhelming them with traffic and rendering them inaccessible to healthcare providers and patients. Such attacks can disrupt online services, hinder communication, and compromise the availability of critical healthcare resources.

- **Insider threats:** The healthcare sector is susceptible to insider threats, where employees with authorized access may intentionally or unintentionally compromise data security. This can involve unauthorized access to patient records, the intentional manipulation of medical data, or the theft of sensitive information.

- **Social engineering and phishing**: Cybercriminals may employ social engineering techniques, such as phishing emails or phone scams, to trick healthcare staff into disclosing sensitive information or granting access to systems. This can result in unauthorized access to healthcare networks, data breaches, or the introduction of malware.

To mitigate the risks and protect the healthcare and public health sectors from cyberattacks, robust cybersecurity measures are crucial. These include implementing secure network infrastructure and training healthcare personnel on cybersecurity best practices.

Information technology sector

The **information technology (IT) sector** encompasses a wide range of industries involved in the development, implementation, and maintenance of computer systems, software, networks, and digital services. It is a crucial sector that drives innovation, enables communication, and supports various sectors of the economy. IT services include software development, network administration, cybersecurity, data management, cloud computing, and technical support.

Impact of a compromised information technology sector

If the IT sector were compromised or under attack, it could have far-reaching consequences impacting businesses, governments, and individuals. Some potential impacts include the following:

- **Disruption of business operations**: Attacks on IT systems can disrupt business operations, leading to downtime, loss of productivity, and financial losses. This can affect organizations of all sizes, from small businesses to large corporations, impacting their ability to serve customers, deliver products and services, and conduct day-to-day operations.

- **Data breaches and information theft**: The IT sector handles vast amounts of sensitive data, including customer information, financial records, and intellectual property. A cyberattack can result in data breaches, where sensitive data is stolen or exposed. This can have severe consequences, including financial fraud, identity theft, reputational damage, and legal and regulatory penalties.

- **Compromised CI**: Attacks on IT systems can target CI such as power grids, transportation systems, telecommunications networks, and healthcare facilities. Compromising these systems can lead to service disruptions, loss of control, and potential safety risks for individuals and communities.

- **Intellectual property theft**: The IT sector is a prime target for intellectual property theft, where attackers seek to steal valuable information, trade secrets, or proprietary software code. This can result in financial losses, loss of competitive advantage, and hindered innovation and technological advancements.

- **Cyber espionage and state-sponsored attacks**: Nation-states may conduct cyber espionage or launch targeted attacks on IT systems to gain access to classified information, government secrets, or sensitive corporate data. These attacks can have significant geopolitical implications, impacting national security and economic stability.

In conclusion, a compromise of the IT sector poses serious risks to businesses, governments, and individuals, including disruption of operations, data breaches, compromised CI, intellectual property theft, and cyber espionage.

Cyberattack scenarios in the information technology sector

Several cyberattack scenarios that pose risks to the IT sector include the following:

- **Malware attacks**: Malicious software, such as viruses, worms, or ransomware, can infiltrate IT systems, compromise network security, and disrupt operations. This can result in data loss, system corruption, or unauthorized access to sensitive information.

- **DDoS attacks**: Attackers may launch DDoS attacks on IT infrastructure, overwhelming networks or servers with massive amounts of traffic, rendering them inaccessible to legitimate users. These attacks can lead to service disruptions, financial losses, and reputational damage.

- **Phishing and social engineering**: Cybercriminals often employ phishing techniques to deceive users into revealing sensitive information, such as passwords or financial details. Social engineering tactics can manipulate individuals into performing actions that compromise IT security, such as clicking on malicious links or downloading malware-infected files.

- **Zero-day exploits**: Zero-day vulnerabilities refer to unknown security flaws in software or systems that attackers exploit before developers can patch them. These exploits can enable attackers to gain unauthorized access, steal data, or compromise systems without detection.

- **Insider threats**: Insider threats involve employees or authorized individuals who misuse their access privileges to compromise IT systems. This can include theft of sensitive data, sabotage of IT infrastructure, or unauthorized disclosure of confidential information.

To mitigate the risks and protect the IT sector from cyberattacks, organizations must prioritize cybersecurity measures. These include implementing robust firewalls and intrusion detection systems, regularly updating software and systems, conducting employee training on cybersecurity best practices, implementing multi-factor authentication, and performing regular security audits and vulnerability assessments.

Nuclear reactors, materials, and waste sector

The **nuclear reactor sector** plays a crucial role in providing a significant portion of the world's electricity through nuclear power generation. It involves the operation and maintenance of nuclear power plants, which harness the energy released from nuclear reactions to produce electricity. This sector requires stringent safety measures and regulatory oversight due to the potential risks associated with nuclear technology.

Impact of a compromised nuclear reactor sector

If the nuclear reactor sector were compromised or under attack, it could have severe consequences on various levels. Here are some potential impacts:

- **Safety risks and radioactive release**: Attacks on nuclear reactors can result in safety breaches, leading to the release of radioactive materials into the environment. This poses a significant risk to public health and the environment, as exposure to radiation can cause serious health effects, including cancer and genetic damage.

- **Power disruption and energy shortages**: Compromised nuclear reactors may require shutdown or reduced power output for safety reasons. This can lead to power disruptions and energy shortages, affecting the reliability of the electricity supply to homes, businesses, and CI. The loss of nuclear power generation capacity may also strain the existing energy infrastructure and result in increased reliance on other energy sources.

- **Environmental contamination**: A cyberattack on the nuclear reactor sector could potentially target the control systems, causing malfunctions or errors that result in environmental contamination. Contaminated soil, water, or air in the vicinity of the reactors can have long-term ecological consequences and require extensive cleanup efforts.

- **Damage to infrastructure**: Cyberattacks on CI components of nuclear reactors, such as cooling systems or emergency response systems, could lead to physical damage and operational disruptions. This can impede the safe operation of the reactors, potentially exacerbating safety risks and prolonging recovery efforts.

Cyberattack scenarios in the nuclear reactor sector

Several cyberattack scenarios pose risks to the nuclear reactor sector:

- **Stuxnet-like attack**: A sophisticated attack similar to the Stuxnet worm, specifically designed to target the control systems of nuclear reactors, could disrupt or manipulate critical processes, compromising safety mechanisms and potentially causing operational failures.

- **Malware infection**: Cybercriminals could target the IT infrastructure and personnel of nuclear reactors, aiming to introduce malware into the systems. This malware may disrupt operations, compromise control systems, or facilitate unauthorized access to CI.

- **Phishing and social engineering**: Attackers may employ phishing techniques or social engineering tactics to deceive employees working in the nuclear reactor sector. By tricking them into revealing sensitive information or gaining unauthorized access to systems, attackers can compromise the security of the reactors and associated infrastructure.

- **Insider threats**: Insider threats from disgruntled employees or individuals with malicious intent within the nuclear reactor sector pose significant risks. Insiders with access to critical systems or sensitive information could intentionally sabotage operations or facilitate external attacks.

- **Supply chain compromise**: The complex supply chains supporting the nuclear reactor sector are potential targets for cyberattacks. By compromising suppliers or introducing malicious components, attackers can infiltrate the sector's infrastructure and gain unauthorized access to critical systems.

To safeguard the nuclear reactor sector against cyberattacks, robust cybersecurity measures are essential. These include implementing strict access controls, conducting regular security assessments, employing advanced intrusion detection and prevention systems, ensuring secure supply chains, educating personnel about cyber threats and best practices, and collaborating with governmental agencies and international organizations to share threat intelligence and strengthen cybersecurity defenses. The nuclear industry also operates under strict regulations and safety protocols to mitigate risks and maintain the highest levels of safety and security.

Transportation system sector

The **transportation system sector** encompasses various modes of transportation, including air, land, and sea, and plays a critical role in enabling the movement of people and goods across regions and countries. It includes infrastructure such as airports, seaports, railways, highways, and public transportation systems. The sector relies heavily on complex networks, information systems, and technology to ensure efficient and safe transportation operations.

Impact of a compromised transportation system sector

If the transportation system sector were compromised or under attack, it could have far-reaching consequences affecting both individuals and economies. Here are some potential impacts:

- **Disruption of services**: Attacks on transportation systems can lead to widespread disruptions, delays, and cancellations of flights, train services, or maritime operations. This can cause significant inconvenience for travelers, logistical challenges for businesses, and economic losses due to interrupted supply chains.

- **Safety risks**: Compromised transportation systems can pose significant safety risks. For example, attacks targeting air traffic control systems could disrupt the communication and coordination of aircraft, potentially leading to accidents or collisions. Attacks on railway systems could affect signaling and control systems, jeopardizing train operations and passenger safety.

- **Economic impact**: The transportation system sector is a vital component of global trade and economic activity. Disruptions or attacks on transportation infrastructure can result in economic losses due to reduced productivity, increased transportation costs, and decreased tourism and business activities. This can have ripple effects across multiple industries and sectors.

- **Public confidence and trust**: A compromised transportation system can erode public confidence and trust in the reliability and security of transportation services. Travelers and businesses may become hesitant to utilize the transportation system, leading to decreased passenger numbers and reduced economic activity.

In conclusion, a compromise of the transportation system sector can have wide-ranging impacts, including service disruptions, safety risks, economic consequences, and a loss of public confidence. Safeguarding the transportation infrastructure is crucial to ensure the smooth functioning of travel, trade, and overall economic stability.

Cyberattack scenarios in the transportation system sector

Several cyberattack scenarios pose risks to the transportation system sector:

- **Ransomware attacks**: Cybercriminals may target transportation agencies or organizations with ransomware, encrypting critical systems or data and demanding a ransom for their release. This can paralyze operations and hinder the ability to provide services until the ransom is paid or the systems are restored.

- **Control system manipulation**: Attackers may attempt to manipulate or disrupt control systems governing transportation infrastructure, such as traffic management systems, air traffic control systems, or railway signaling systems. By exploiting vulnerabilities in these systems, they can cause chaos, delays, or even accidents.

- **GPS spoofing**: **Global Positioning System** (GPS) spoofing involves sending false signals to manipulate the location or timing information received by transportation vehicles or systems. By spoofing GPS signals, attackers can misguide navigation systems, leading to incorrect routes, collisions, or intentional misdirection of transportation assets.

- **Unauthorized access to transportation systems**: Attackers targeting transportation systems may seek unauthorized access to critical systems, such as ticketing or reservation databases, passenger information systems, or control interfaces. This can result in data breaches, identity theft, or unauthorized manipulation of passenger records or travel itineraries.

- **Infrastructure targeting**: The physical infrastructure of transportation systems, such as bridges, tunnels, or key transportation hubs, could be targeted for cyberattacks. By compromising the operational systems or infrastructure components, attackers can disrupt transportation flow, compromise structural integrity, or facilitate physical attacks.

To mitigate the risks of cyberattacks in the transportation system sector, robust cybersecurity measures are crucial. This includes implementing strong access controls, network segmentation, intrusion detection systems, and encryption mechanisms. Regular security assessments, employee training on cybersecurity best practices, and information-sharing collaborations with industry partners and government agencies are also vital for maintaining the resilience and security of the transportation system sector.

Water and wastewater sector

The **water and wastewater sector** plays a critical role in providing clean and safe water for drinking, industrial use, and sanitation purposes. It encompasses various entities such as water treatment plants, distribution systems, wastewater treatment facilities, and water supply infrastructure. The sector is responsible for collecting, treating, and supplying water to communities and ensuring the proper management of wastewater.

Impact of a compromised water and wastewater sector

If the water and wastewater sector were compromised or under attack, it could have severe consequences for public health, the environment, and economic stability. Here are some potential impacts:

- **Public health risks**: A compromised water and wastewater sector can pose significant risks to public health. Water supply systems may be targeted to contaminate drinking water with harmful substances, pathogens, or chemicals. This can lead to widespread illnesses, outbreaks of waterborne diseases, and potential loss of life.

- **Environmental damage**: Attacks on the water and wastewater sector can result in environmental damage. For example, tampering with wastewater treatment systems can lead to the release of untreated or inadequately treated wastewater into rivers, lakes, or oceans, causing pollution and harming aquatic ecosystems. Contamination of water sources can have long-lasting ecological effects.

- **Disruption of services**: Attacks on the water and wastewater sector can disrupt the supply of clean water to communities. This can lead to water shortages, reduced water quality, and interruptions in essential services such as drinking water, sanitation, and firefighting. Communities may face difficulties in meeting basic needs and maintaining hygiene standards.

- **Economic impact**: Compromised water and wastewater systems can have significant economic implications. Industries that rely on a stable and reliable water supply, such as agriculture, manufacturing, and energy production, may face disruptions in their operations. Economic productivity can decline, and communities dependent on water-related tourism may experience negative impacts.

In conclusion, a compromise of the water and wastewater sector not only poses serious risks to public health and the environment but also has far-reaching consequences for economic stability and various industries dependent on a reliable water supply.

Cyberattack scenarios in the water and wastewater sector

Several cyberattack scenarios pose risks to the water and wastewater sector:

- **Infrastructure disruption**: Attackers may target the operational systems and control networks of water treatment plants, pumping stations, or wastewater treatment facilities. By gaining unauthorized access or exploiting vulnerabilities, they can disrupt critical processes, control mechanisms, or remote monitoring systems, leading to service interruptions or compromised water quality.

- **Data manipulation and theft**: Cybercriminals may attempt to manipulate data within water management systems, including water quality monitoring data or billing systems. Manipulating data can misrepresent water quality levels, hinder accurate decision-making, or facilitate fraudulent activities.

- **Phishing and social engineering**: Attackers may employ phishing emails, social engineering techniques, or targeted spear-phishing campaigns to gain unauthorized access to the network infrastructure or internal systems of water and wastewater organizations. Once inside the network, they can exploit vulnerabilities, escalate privileges, or launch further attacks.

- **DDoS attacks**: Water and wastewater systems can be targeted with DDoS attacks, overwhelming network resources, control systems, or communication channels. These attacks can disrupt operations, compromise system availability, and hinder the ability to monitor and respond to critical events.

- **Insider threats**: Insiders with authorized access to water and wastewater systems can misuse their privileges or engage in malicious activities. This can include intentionally tampering with control systems, sabotaging processes, or leaking sensitive information.

To protect the water and wastewater sector from cyberattacks, robust cybersecurity measures are essential. This includes implementing secure network architectures, access controls, encryption mechanisms, and intrusion detection systems. Regular vulnerability assessments, staff training on cybersecurity best practices, and collaborations with cybersecurity experts and government agencies are crucial for maintaining the resilience and security of the water and wastewater sector. Additionally, establishing incident response plans and conducting regular exercises to test the response capabilities can help minimize the impact of potential cyber incidents.

Summary

In this chapter, we explored the fundamental concepts of CI and its significance in our society. You now understand what it entails, comprising 16 sectors crucial to the United States, including examples such as the electrical grid, the chemical industry, and commercial facilities.

Moreover, you've gained insight into the importance of safeguarding CI. These sectors aren't just vital for national security; they're integral to economic stability, public health, and safety.

You can identify and categorize various CI sectors, recognize their vital roles in our daily lives, and comprehend the far-reaching consequences of compromising CI, impacting not only specific sectors but also the nation as a whole.

You've been exposed to various case scenarios stemming from cyberattacks on CI, enabling you to envision real-world implications, and you can analyze and assess risks linked to vulnerabilities within these sectors, contributing to informed decision-making and mitigation strategies.

As you continue through this book, these foundational lessons and skills will serve as a solid basis for exploring the challenges, solutions, and complexities of protecting our nation's critical infrastructure in depth. Prepare to explore a myriad of topics that will empower you to contribute to the security and resilience of the vital systems underpinning our society.

In the upcoming chapter, we will explore the escalating risks of cyberattacks on CI. You'll gain insights into the vulnerabilities of our interconnected systems and the imperative of bolstering defenses.

References

To learn more about the topics that were covered in this chapter, take a look at the following resources:

- *Cybersecurity and Infrastructure Security Agency (CISA). (n.d.). Critical Infrastructure Sectors*: `https://www.cisa.gov/critical-infrastructure-sectors`

- *U.S. Department of Homeland Security. (n.d.). Chemical Sector*: `https://www.dhs.gov/chemical-sector`

- *U.S. Department of Homeland Security. (n.d.). Commercial Facilities Sector*: `https://www.dhs.gov/commercial-facilities-sector`

- *Cybersecurity and Infrastructure Security Agency (CISA). (n.d.). Communications Sector*: `https://www.cisa.gov/communications-sector`

- *U.S. Department of Homeland Security. (n.d.). Critical Manufacturing Sector*: `https://www.dhs.gov/critical-manufacturing-sector`

- *National Infrastructure Protection Plan Sector-Specific Plans*: `https://www.dhs.gov/xlibrary/assets/nipp_sctrplans.pdf`

- *Chemical Sector Cybersecurity Framework Implementation Guidance*: `https://www.cisa.gov/sites/default/files/publications/Chemical_Sector_Cybersecurity_Framework_Implementation_Guidance_FINAL_508.pdf`

- *United States Government Accountability Office. (2020). Critical Infrastructure Protection*: `https://www.gao.gov/assets/gao-20-424.pdf`

- *U.S. Government Accountability Office. (2020, Feb). Critical Infrastructure Protection: Additional Actions Needed to Identify Framework Adoption and Resulting Improvements. GAO*: `https://www.gao.gov/products/gao-20-299`

- *U.S. Department of Homeland Security. (2021, March 17). Commercial Facilities Sector*: `https://www.cisa.gov/commercial-facilities-sector`

- *Secure Cyberspace and Critical Infrastructure*: `https://www.dhs.gov/secure-cyberspace-and-critical-infrastructure`

- *U.S. Department of Homeland Security. (2015). Commercial facilities sector-specific plan: An annex to the national infrastructure protection plan. DHS*: `https://www.cisa.gov/sites/default/files/publications/nipp-ssp-commercial-facilities-2015-508.pdf`

2

The Growing Threat of Cyberattacks on Critical Infrastructure

In the modern era, a new wave of transformation has been ushered in by the emergence of intricate and sophisticated information infrastructures, comprising global computer networks and highly developed control systems. Our existing infrastructures have reached unprecedented levels of performance, yielding remarkable and often awe-inspiring outcomes. The profound impact of these technological advancements extends beyond infrastructure alone, permeating our entire culture and shaping the very fabric of our society.

Yet, as we rejoice in the triumphs of our progress, we must also remain vigilant about the potential risks that accompany our reliance on these technologies. Our increasing dependence on interconnected systems has introduced hidden vulnerabilities, leaving us exposed to both natural calamities and human-made disasters such as cyberattacks. Safeguarding against these potential threats requires careful planning, collaboration, and foresight to ensure that as we continue to strengthen our national capacity and economic prowess, we do so with an eye toward resilience and preparedness for any eventuality.

In this chapter, you will learn about the following:

- The history of cyberattacks on CI
- An analysis of the current global situation
- National cybersecurity strategies

A brief history of CI protection and attacks

In the ever-evolving landscape of modern society, the protection of **critical infrastructure (CI)**has become crucial. Let's overview the historical development of CI protection and the evolving nature of the attacks that have threatened these vital systems.

The impact of the 9/11 attacks on CI

Prior to September 11, the destruction of CI was a concern that received relatively little attention in the public consciousness. While it was recognized among certain government agencies, security experts, and specialized industries, the general public often remained unaware of the potential vulnerabilities and consequences associated with such attacks.

The prevailing view was that CI, such as power plants, transportation systems, communication networks, and key government facilities, were well-protected and less susceptible to large-scale damage. Attention was primarily focused on traditional security threats, such as armed conflicts between nations or localized acts of violence.

However, the events of September 11, 2001, dramatically shifted this perspective. The coordinated terrorist attacks on the World Trade Center in New York City and the Pentagon in Washington, D.C. demonstrated the immense impact that an assault on CI could have on a nation. These attacks, carried out with hijacked commercial airplanes, not only resulted in the tragic loss of life but also caused extensive damage to iconic structures and disrupted vital services, severely affecting the nation's economy and sense of security.

In the aftermath of 9/11, there was a paradigm shift in how CI protection was perceived and approached. Governments worldwide began to reassess and reinforce security measures to safeguard key facilities and systems from potential threats. The focus shifted toward comprehensive risk assessments, investing in more robust security protocols, and enhancing coordination among various agencies and industries responsible for maintaining CI.

The events of September 11 served as a wake-up call, prompting the recognition that the destruction of CI could have far-reaching consequences on national security, public safety, and economic stability. Consequently, efforts to protect and secure these vital assets have since become a primary priority for governments and organizations around the globe. While challenges remain, the heightened awareness of CI's vulnerability and the collective determination to bolster its protection have significantly evolved since that pivotal day.

The twin beams of light in NYC, known as the *Tribute in Light*, represent a poignant and solemn tribute to the tragic events of 9/11. These beams of light serve as a symbol of remembrance for the lives lost during the September 11, 2001 terrorist attacks and the enduring spirit of resilience in the face of adversity. They are a powerful and iconic symbol of unity and the collective strength of the American people in the aftermath of the tragedy:

Figure 2.1 – Symbolizing resilience and remembrance: Twin beams of light
in NYC's skyline after 9/11 (Source: Lerone Pieters on Unsplash)

Same old attacks throughout history

The sabotage of CI is not a new phenomenon. It has been a tactic employed throughout history to disrupt and weaken civilizations. In ancient history, we can find examples of attacks on CI that had significant consequences on societies of that time.

Battle of Salamis – 480 BCE

An example from ancient history is the Battle of Salamis in 480 BCE during the Greco-Persian Wars. The Persian King Xerxes sought to invade Greece and to achieve this, his fleet was heavily reliant on a strategic natural waterway known as the Strait of Salamis. The Greek naval forces, led by Themistocles, devised a cunning plan to exploit this vulnerability. They lured the Persian fleet into the narrow and congested waters of the strait, where the larger Persian ships struggled to maneuver effectively. This tactical move led to the destruction of a significant portion of the Persian fleet, disrupting their invasion plans and safeguarding Greek independence. The strategic map of the Battle of Salamis emerges as a key battlefield blueprint, shaping the course of ancient naval warfare. Explore its intricacies and significance in the following figure:

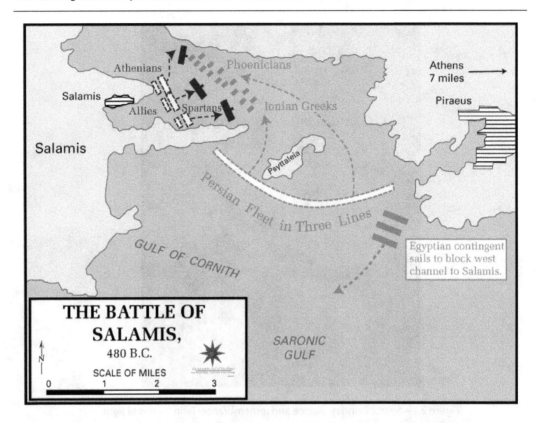

Figure 2.2 – The Battle of Salamis (Source: The Department of History, United States Military Academy)

Operation Chastise – 1943

In more recent times, the German military's Operation Chastise in 1943 during World War II exemplifies the sabotage of CI. The operation aimed to cripple industrial production and impede Nazi Germany's war efforts. The target was a series of dams in the Ruhr Valley. The British Royal Air Force's 617 Squadron, led by Wing Commander Guy Gibson, used innovative bouncing bombs specifically designed to breach the dams' walls. The destruction of these key water reservoirs severely hampered industrial production and disrupted vital transportation and energy supply routes in the region.

Figure 2.3 – Monoplanes Squadron in WWII (Source: Museums Victoria on Unsplash)

Stuxnet cyberattack – 2010

The Stuxnet cyberattack in 2010 stands out as a landmark event in the history of CI sabotage. Stuxnet was a highly sophisticated computer worm that specifically targeted Iran's nuclear facilities, particularly its uranium enrichment centrifuges. The attackers utilized advanced cyber capabilities to infiltrate the computer systems controlling the centrifuges, causing them to malfunction and sabotage Iran's nuclear program. This attack demonstrated the immense power of cyberwarfare in disrupting CI, showcasing that modern technological advancements have opened new avenues for such threats.

Ukraine's power grid cyberattack – 2015

As we progress into the modern era, cyberwarfare has emerged as a potent tool for targeting CI. The 2015 cyberattack on Ukraine's power grid serves as a striking example. In this incident, sophisticated hackers breached the control systems of several energy distribution companies, causing widespread power outages in Ukraine. The attack effectively disrupted CI, leaving thousands without electricity during the harsh winter. This event highlighted the vulnerability of interconnected systems and the potential consequences of cyberattacks on essential infrastructure.

Baltimore cyberattack, Maryland, USA – 2019

Additionally, the 2019 cyberattack on the city of Baltimore, Maryland, USA, serves as a modern example of CI sabotage. A ransomware attack targeted the city's computer systems, paralyzing critical services and infrastructure, including email communication, payment systems, and public services. The attack had a significant impact on the city's operations, highlighting the growing risk of cyber threats on vital urban infrastructure.

These historical and modern examples demonstrate that the sabotage of CI has persisted throughout the ages. Whether through strategic military tactics, targeted bombing campaigns, or cyberwarfare, the vulnerability of essential systems remains a constant concern for societies worldwide. The protection of CI requires ongoing vigilance, collaboration among stakeholders, and the implementation of robust security measures to safeguard against potential threats and ensure the stability and resilience of nations and cities.

Executive order 13010

Executive Order 13010, issued by President Bill Clinton on July 15, 1996, was a landmark step in the United States' efforts to protect and strengthen its CI. The order was titled **Critical Infrastructure Protection** and laid the groundwork for enhancing the security and resilience of essential systems against potential threats.

One of the significant aspects of Executive Order 13010 was the establishment of the **President's Commission on Critical Infrastructure Protection (PCCIP)**. The PCCIP was tasked with conducting a comprehensive review and assessment of the nation's CI, identifying vulnerabilities, and proposing strategies to mitigate risks. This commission brought together experts from various government agencies, private sector industries, and academic institutions to collaborate on a unified approach to safeguarding CI.

A crucial contribution of the Executive Order was the first formal definition of *critical infrastructure*. In this order, CI was defined as *"systems and assets, whether physical or virtual, so vital to the United States that the incapacity or destruction of such systems and assets would have a debilitating impact on security, national economic security, national public health or safety, or any combination of those matters."*

This definition encompassed a wide range of sectors, including telecommunications, energy, transportation, finance, water supply, emergency services, and information technology. By categorizing these systems and assets as CI, the order recognized their indispensable role in the functioning of society and the nation's overall security.

Executive Order 13010 emphasized the importance of collaboration between the government and the private sector in protecting CI. It encouraged the sharing of information and expertise to develop effective strategies and response plans to potential threats.

The order directed federal agencies to assess the vulnerabilities of CI within their respective areas of responsibility and develop plans for improving security and resilience. It also outlined the establishment of sector-specific plans to address the unique challenges faced by different CI sectors.

Executive Order 13010 represented a significant step toward strengthening the United States' CI protection efforts. By providing a comprehensive definition of CI and fostering collaboration between government agencies and private industries, the order laid the foundation for a more cohesive and proactive approach to safeguarding essential systems from emerging threats. Subsequent administrations have built upon this foundation, reinforcing the nation's commitment to the security and resilience of CI in an ever-changing and interconnected world.

Evolution of a nation's CI protection posture

Over time, a nation's attitude and actions to protect itself from CI attacks have evolved significantly due to the changing nature of threats and advancements in technology. As societies have become more interconnected and reliant on complex infrastructures, safeguarding these vital systems has become of utmost importance for national security and economic stability.

In the past, the protection of CI was often a secondary concern, with a greater focus on traditional military defenses. However, as history showcased the devastating impact of targeted attacks on essential systems, nations began to realize the strategic importance of safeguarding their infrastructures.

During World War II, the devastation caused by attacks on CI, such as power plants, transportation networks, and communication centers, prompted governments to take proactive measures. The establishment of civil defense programs and the strengthening of infrastructure resilience became essential components of national security strategies.

As the world entered the digital age, a new frontier of threats emerged in the form of cyberwarfare. The increasing reliance on computer systems and networks made CI vulnerable to cyberattacks. This evolution led to the rise of cybersecurity as a crucial aspect of protecting CI. Governments invested in developing specialized cyber units, implementing robust cybersecurity measures, and fostering international cooperation to combat cyber threats.

In the modern era, the concept of CI protection has expanded to include not just physical assets but also data and information systems. Nations have recognized the need to safeguard against both physical and virtual threats to ensure comprehensive security.

Collaborative efforts between governments, private sector entities, and international organizations have become more prevalent. Public-private partnerships, such as the **Cybersecurity and Infrastructure Security Agency (CISA)** and Microsoft, play a significant role in sharing information, expertise, and resources to enhance the security of CI. Countries engage in bilateral and multilateral agreements to foster cooperation and coordinate responses to potential attacks.

The use of advanced technologies such as artificial intelligence, machine learning, and big data analytics could also revolutionize CI protection. These technologies enable real-time monitoring, threat detection, and predictive analysis, enhancing a nation's ability to respond swiftly and effectively to potential attacks.

As threats continue to evolve, a nation's attitude toward CI protection must remain dynamic and adaptable. Continuous investment in research, innovation, and training is essential to stay ahead of sophisticated adversaries. Emphasizing resilience, redundancy, and contingency planning is crucial to ensure that CI can withstand and recover from potential attacks.

The evolution of a nation's attitude and actions to protect CI reflects a growing awareness of the interconnectedness of modern societies and the need to proactively defend against emerging threats. By recognizing the strategic importance of safeguarding essential systems and fostering cooperation at all levels, nations can better ensure the security and stability of their citizens and economies.

Evolution of cyberattacks and countermeasures

The history of CI and cyberattacks is closely intertwined with the evolution of technology and society. The increasing reliance on digital technologies and interconnected systems has made CI susceptible to cyber threats and attacks.

The earliest forms of CI were basic and relied on manual processes. Ancient civilizations developed rudimentary systems for transportation, communication, and defense, such as roads, messengers, and fortified structures. As societies advanced, so did their infrastructure needs. The Industrial Revolution marked a turning point, leading to the development of complex energy grids, factories, and communication networks, transforming the way people lived and worked.

However, with the advent of computers and the internet in the latter half of the 20th century, CI underwent a profound transformation. These digital technologies brought unprecedented connectivity, enabling more efficient management, monitoring, and control of critical systems. While this brought immense benefits, it also introduced new vulnerabilities that malicious actors could exploit.

As digital infrastructures became more prevalent, cyberattackers recognized the potential for disruption and chaos. Early cyber incidents in the 1980s and 1990s, driven by curiosity and not necessarily malicious intent, were the precursors to the more sophisticated and targeted cyberattacks that emerged later. The landscape shifted when the Stuxnet malware was discovered in 2010. This highly complex cyberweapon specifically targeted Iran's nuclear program, demonstrating that cyberattacks could have tangible physical consequences on CI.

Subsequent years saw a rise in nation-state attacks on CI, with state-sponsored actors targeting energy, financial institutions, and communication systems to achieve geopolitical and economic goals. The 2015 and 2016 attacks on Ukraine's power grid and the widespread deployment of ransomware, such as WannaCry and NotPetya in 2017, further emphasized the vulnerability of essential systems to cyber threats.

As cyberattacks became more sophisticated, protecting CI required a concerted effort from governments, organizations, and cybersecurity experts. Nations developed national cybersecurity strategies, recognizing the need for comprehensive approaches to safeguarding critical systems. Public-private partnerships emerged as a vital means of enhancing cybersecurity resilience, with collaborative information sharing and coordinated incident response efforts.

To combat evolving threats, organizations adopted robust cybersecurity measures, including regular vulnerability checks, network monitoring, and incident response planning. Employee education on cybersecurity and human error reduction also became key. Advanced technology implementation, such as enhanced network segmentation, strengthened defenses.

Collaboration and information sharing among industry, government, and cybersecurity bodies became essential in fighting cyber threats to infrastructure. This synergy ensures swift threat intelligence sharing and best practice dissemination.

In summary, the interplay between technological growth and threat evolution highlights the escalating need to protect CI from cyberattacks in an increasingly digital world.

Let's explore the key milestones in the history of CI cyberattacks and the corresponding efforts to protect CI:

	Attacks	**Countermeasures**
Early Cyber Incidents (1980s–1990s)	In the early days of the internet, cyberattacks were relatively simple and often driven by curiosity rather than malicious intent. During this period, security breaches were mostly focused on academic and government institutions. Notably, the **Morris Worm** in 1988 is considered one of the first significant cyberattacks. It infected thousands of computers, causing widespread disruption.	As the internet became more prevalent, CI sectors gradually adopted digital technologies for improved efficiency and management. However, cybersecurity measures during this period were relatively basic and focused on traditional firewalls, antivirus software, and perimeter-based defenses. As attacks increased in frequency and sophistication, it became evident that more robust strategies were needed.

	Attacks	Countermeasures
Stuxnet (2010)	Stuxnet was a groundbreaking cyberweapon discovered in 2010. It specifically targeted Iran's nuclear program, aiming to disrupt and destroy centrifuges used for uranium enrichment. Stuxnet marked a significant shift as it demonstrated that cyberattacks could have real-world, physical consequences by targeting CI.	The discovery of Stuxnet in 2010 highlighted the potential consequences of cyberattacks on CI. This sophisticated malware specifically targeted **industrial control systems (ICS)** and brought attention to the need for stronger protections against such attacks. Governments and organizations started to recognize the significance of securing CI from cyber threats and began investing in more advanced cybersecurity solutions.
Rise of Nation-State Attacks (2010s)	In the 2010s, cyberattacks became more sophisticated, and state-sponsored actors began playing a more significant role. Notable attacks during this period include the **Russian 22 days long cyberattack on Estonia in 2007**, the **Iranian attack on Saudi Aramco in 2012**, and the **North Korean attack on Sony Pictures in 2014**. These attacks targeted various CI sectors, highlighting the vulnerability of vital systems to cyber threats.	In response to the escalating cyber threat landscape, many countries developed comprehensive national cybersecurity strategies. These strategies aimed to protect CI by promoting information sharing, collaboration between the public and private sectors, and the establishment of dedicated cybersecurity agencies. These initiatives sought to enhance the overall resilience of CI against cyberattacks.
Ukraine Power Grid Attack (2015 and 2016)	In December 2015 and December 2016, Ukraine experienced two separate cyberattacks on its power grid. The attackers were able to gain control of critical systems, resulting in widespread power outages. These attacks were attributed to Russian state-sponsored hackers and served as a wake-up call for the potential implications of cyberattacks on CI.	Governments and CI operators recognized the importance of collaboration in addressing cyber threats effectively. Public-private partnerships were forged to facilitate information exchange, threat intelligence sharing, and joint cybersecurity exercises. Such collaborations helped bridge the gap between governmental knowledge and private sector expertise, leading to more robust defense measures.

	Attacks	Countermeasures
WannaCry Ransomware (2017)	WannaCry was a global ransomware attack that occurred in May 2017. It targeted computers running Microsoft Windows, encrypting their data and demanding ransom payments in Bitcoin. It significantly impacted various sectors, including healthcare, transportation, and financial services, underscoring the interconnectivity and vulnerability of CI to cyber threats.	The prevalence of ransomware attacks targeting CI increased in the 2010s. Notable incidents, such as WannaCry and NotPetya, brought attention to the importance of data backup, disaster recovery plans, regular software patching, and vulnerability management to prevent and recover from ransomware attacks. Organizations also began investing in cybersecurity training and awareness programs to reduce the likelihood of successful phishing attacks and malware infections.
Triton/Trisis (2017)	The Triton/Trisis attack targeted a petrochemical plant in Saudi Arabia in 2017. It was the first known cyberattack explicitly designed to manipulate ICS and **safety instrumentation systems (SIS)**. The attack sought to cause physical damage and highlighted the potential for cyberattacks to jeopardize human safety in CI facilities.	As cyberattackers increasingly targeted ICS and **operational technology (OT)** environments, a specialized focus on ICS security emerged. Organizations and cybersecurity experts started developing solutions specifically tailored to protect these critical systems from cyber threats. This included securing legacy systems, implementing network segmentation, and integrating cybersecurity into the design of new ICS infrastructure.
SolarWinds (2019)	The SolarWinds Orion software supply chain attack was one of the most sophisticated cyberattacks in history. The attack infected over 18,000 organizations, including government agencies, businesses, and universities.	The incident exposed the vulnerabilities inherent to software supply chains, prompting a thorough review of vendor security practices. Companies began implementing more stringent vetting procedures for third-party software providers and regularly assessing their security controls. Zero trust architecture gained traction as a robust approach to prevent lateral movement within networks, ensuring that no device or user is inherently trusted.

	Attacks	Countermeasures
Colonial Pipeline ransomware attack (2021)	In May 2021, the Colonial Pipeline, a major U.S. fuel pipeline, experienced a ransomware attack by the DarkSide cybercriminal group. This attack disrupted the pipeline's operations, leading to widespread fuel shortages and panic buying on the East Coast. The attack highlighted the vulnerability of CI to cyber threats and the cascading effects such incidents can have on society and the economy.	In the aftermath, there was a heightened focus on strengthening cybersecurity defenses across various sectors, particularly in CI. This included enhancing network security, adopting advanced threat detection tools, and conducting rigorous security assessments. The incident also spurred improved collaboration between the public and private sectors for sharing threat intelligence and coordinating incident response efforts. These measures aimed to bolster resilience against future cyber threats and safeguard essential systems.

Table 2.1 – Key milestones: Cyberattacks on CI

The evolution of CI and cyberattacks has followed a dynamic and interconnected path. As society became increasingly reliant on digital technologies, CI emerged as a prime target for cyber adversaries seeking to disrupt operations, cause harm, or extract ransom payments. Over the years, cyberattacks on CI have grown in scale, sophistication, and frequency, posing significant challenges to the security and stability of essential systems.

The state of CI in the face of cyberattacks

Cyberattacks on CI have continued to escalate in both frequency and complexity. There have been several notable incidents affecting power grids, transportation networks, and healthcare systems globally. Nation-state actors, criminal organizations, and hacktivists have continued to exploit vulnerabilities, causing significant economic and social repercussions. Let's explore some of the latest cyberattacks and scenarios.

COVID-19-period cyberattack landscape

During the 2020 COVID-19 pandemic, cybercriminals seized the opportunity to target CI organizations that were at the forefront of healthcare, medical research, and vaccine development. The pandemic presented a unique and chaotic environment, with hospitals overwhelmed, research institutions racing to find solutions, and the global population seeking information and support. In this atmosphere of urgency and uncertainty, cybercriminals saw an opportunity to exploit vulnerabilities and launch various cyberattacks.

Phishing campaigns became prevalent during this time, with attackers impersonating legitimate organizations such as the **World Health Organization (WHO)**, the **Centers for Disease Control and Prevention (CDC)**, and other healthcare agencies. They sent deceptive emails masquerading as critical updates or important health information related to the pandemic. These emails contained malicious links or attachments designed to steal sensitive information, compromise systems, or install malware.

Ransomware attacks also surged during the pandemic. Cybercriminals targeted healthcare institutions, research labs, and pharmaceutical companies with ransomware, encrypting their data and demanding exorbitant ransom payments for its release. The urgency of the pandemic meant that many organizations might have been more inclined to pay the ransom to regain access to essential data and services.

Other cyber threats such as **distributed denial of service (DDoS)** attacks, data breaches, and social engineering tactics were on the rise. Cybercriminals took advantage of the distraction caused by the pandemic and the rapid shift to remote work to exploit weaknesses in security infrastructures.

The motivation behind these attacks varied. Some attackers sought financial gain, exploiting the chaos to extort money through ransom payments or by selling stolen medical data on the dark web. Others may have had ideological or political motives, aiming to disrupt healthcare services or vaccine development efforts for their own agenda.

To counter these threats, CI organizations had to strengthen their cybersecurity defenses rapidly. They invested in advanced threat detection and prevention solutions, intrusion detection systems, and user behavior analytics to identify and mitigate potential threats in real time. Additionally, organizations bolstered employee training and awareness programs to educate staff about recognizing phishing attempts and following secure practices while working remotely.

Governments and cybersecurity agencies also played a crucial role in responding to the increased cyber threats. They issued advisories, disseminated threat intelligence, and collaborated with private sector entities to share information and coordinate incident response efforts.

The 2020 pandemic demonstrated that cybercriminals have no qualms about exploiting global crises for their own gain. It also highlighted the critical importance of robust cybersecurity measures and proactive threat mitigation strategies, especially for organizations involved in CI sectors. As the world continues to grapple with the pandemic's aftermath and embraces a more digitally interconnected future, cybersecurity remains an ongoing priority to safeguard essential services and protect sensitive information from evolving cyber threats.

The Colonial Pipeline ransomware attack

The **Colonial Pipeline ransomware attack** in 2021 was a significant cybersecurity incident that had far-reaching implications for the United States. The attack targeted Colonial Pipeline, one of the largest fuel pipeline operators in the country, which operates a critical pipeline system supplying gasoline, diesel, jet fuel, and other refined petroleum products to the East Coast.

The DarkSide ransomware group, believed to have Eastern European origins, was responsible for the attack. They gained unauthorized access to Colonial Pipeline's computer systems and deployed ransomware that encrypted critical data, making it inaccessible to the company. In ransomware attacks, the perpetrators typically demand a ransom payment in exchange for providing the decryption key to restore access to the encrypted data.

In response to the attack and as a precautionary measure, Colonial Pipeline took the decision to shut down its entire pipeline system, spanning over 5,500 miles, to prevent the further spread of the ransomware and protect its operations. This shutdown resulted in an immediate disruption of fuel supplies, causing panic-buying, price spikes, and fuel shortages in various states along the East Coast.

The pipeline shutdown had significant implications for the transportation and availability of gasoline and other petroleum products. Gasoline prices soared, and long lines formed at gas stations as consumers rushed to fill their tanks amid fears of prolonged shortages. Additionally, CI operators and government authorities faced the challenge of ensuring the continued supply of essential fuel to emergency services, hospitals, airports, and other essential services.

The Colonial Pipeline attack served as a stark reminder of the vulnerability of CI to cyber threats. The incident highlighted the potential consequences of cyberattacks on essential systems and their ripple effects on the economy and daily life. It underscored the need for robust cybersecurity measures and proactive threat mitigation strategies for CI operators.

Following the attack, there was increased attention on the importance of implementing measures to prevent and respond to such incidents. Organizations across various sectors, especially CI, accelerated efforts to strengthen their cybersecurity defenses, including enhancing network security, adopting advanced threat detection tools, and conducting rigorous security assessments.

The incident also prompted discussions on the need for improved collaboration between public and private sectors to share threat intelligence and coordinate incident response efforts. Government agencies, cybersecurity firms, and industry stakeholders worked together to investigate the attack, identify the threat actors, and gather actionable information to prevent similar incidents in the future.

In the aftermath of the Colonial Pipeline attack, cybersecurity and protecting CI have become even more prominent on the national security agenda. Policymakers and industry leaders have been working toward enhancing resilience and establishing robust protocols to safeguard essential systems from cyber threats. Continuous vigilance, information sharing, and proactive cybersecurity measures remain vital to safeguarding CI and mitigating the potential impact of future cyberattacks on essential services.

A comprehensive illustration of the components of the Colonial Pipeline cyberattack is shown in the following figure:

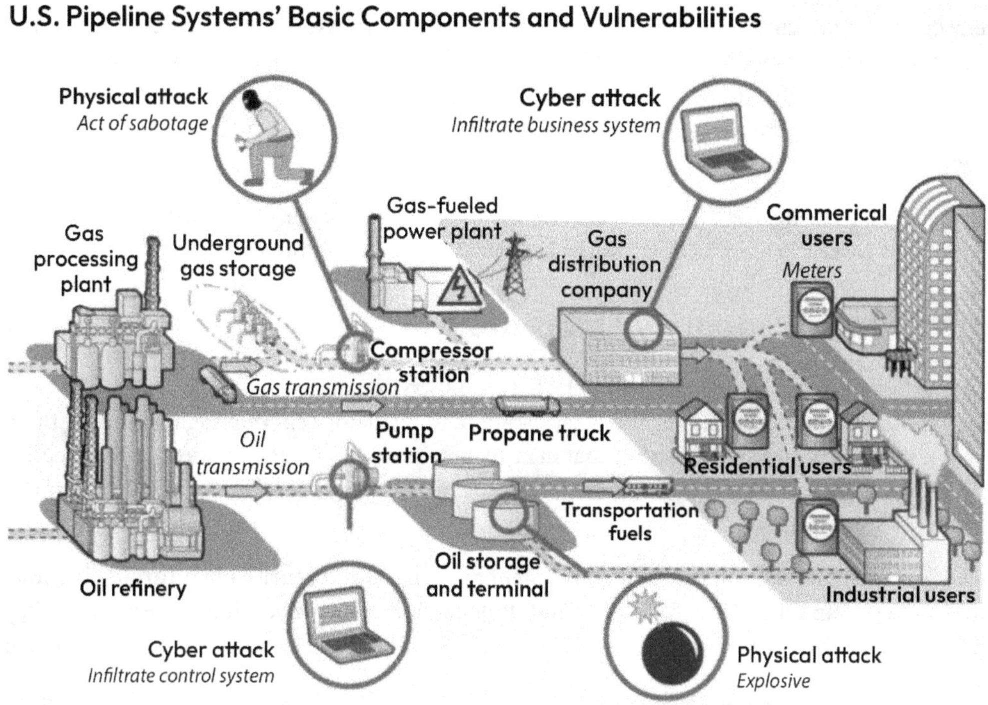

U.S. Pipeline Systems' Basic Components and Vulnerabilities

Figure 2.4 – Colonial Pipeline attack (source: https://www.gao.gov/)

Attacks in 2023

Only in 2023, there have been a number of significant attacks that have highlighted the growing threat to CI:

- March 2023 witnessed the White House's National Cybersecurity Strategy, elevating ransomware to the status of a top-tier national security concern. This decision came in response to a sequence of attacks that impacted critical national infrastructure services, including food suppliers, healthcare institutions, and educational facilities.

- By April 2023, it became known that the group behind a significant breach affecting a VoIP company, 3CX, had also compromised two CI entities in the energy sector. One of these infiltrations occurred within the United States and the other in Europe.

- In May 2023, the UK's **National Cyber Security Centre (NCSC)** issued a warning concerning a novel group of cyber threats with Russian affiliations, which presented a menace to the vital infrastructure of the United Kingdom.

Governments and organizations have increased their focus on cybersecurity measures, information sharing, and incident response planning. However, the dynamic nature of cyber threats means that staying ahead of attackers remains a continuous challenge in protecting CI. As technology continues to advance, it is essential to remain vigilant and adaptive to defend against cyber threats to our most vital systems.

As we move on to the next section, we'll explore the dynamic landscape of **national cybersecurity strategies** that governments are employing to counter these emerging threats.

National cybersecurity strategies

National cybersecurity strategies for CI vary from country to country, as each nation tailors its approach based on its specific needs, threat landscape, and regulatory environment. However, there are some common themes and objectives that many countries address in their national cybersecurity strategies for CI. These strategies aim to enhance the overall resilience of essential systems and protect them from cyber threats.

The United States has taken significant steps to enhance the cybersecurity of CI through various national strategies and initiatives. While specific strategies may evolve over time, here are some key elements and initiatives that have been part of the United States national cybersecurity approach to CI:

- **National Infrastructure Protection Plan (NIPP)**: The NIPP serves as the overarching strategy for CI security and resilience. It outlines the roles and responsibilities of various federal agencies, private sector partners, and state and local governments in safeguarding CI sectors.

- **Presidential Policy Directive 21 (PPD-21)**: Issued in 2013, PPD-21 identifies 16 CI sectors and establishes a framework for collaboration between government and private sector entities to address cybersecurity risks and protect against cyber threats.

- **Cybersecurity Framework**: Developed by the **National Institute of Standards and Technology (NIST)**, the Cybersecurity Framework provides guidelines, best practices, and standards to help CI operators assess and improve their cybersecurity risk management.

- **Department of Homeland Security (DHS) programs**: The DHS plays a central role in coordinating efforts to protect CI. Initiatives such as the **Critical Infrastructure Cyber Community (C³) Voluntary Program** and the **National Cybersecurity and Communications Integration Center (NCCIC)** support information sharing, incident response, and collaborative efforts with private sector partners.

- **Information Sharing and Analysis Centers (ISACs)**: The United States has established ISACs for various CI sectors, allowing private sector organizations to share cyber threat intelligence with each other and with government agencies in real time.

- **Executive order on improving the nation's cybersecurity**: In May 2021, President Joe Biden issued an executive order aimed at enhancing the nation's cybersecurity. The order focuses on improving the federal government's cybersecurity practices, fostering private-sector collaboration, and modernizing cybersecurity defenses.

- **Enhanced incident response**: The United States has developed Cyber Response Playbooks to guide federal agencies in responding to cyber incidents that may impact CI. These playbooks facilitate coordinated response efforts and support collaboration with private sector partners.

- **Public-private partnerships**: The U.S. government emphasizes the importance of public-private partnerships in protecting CI. Initiatives such as the **Enhanced Cybersecurity Services (ECS)** program enable trusted cybersecurity providers to share threat intelligence with private sector entities.

- **Continuous Diagnostics and Mitigation (CDM) Program**: The CDM Program is a cybersecurity initiative led by CISA within the U.S. DHS. The program aims to improve the overall cybersecurity posture of federal government agencies by providing them with tools, resources, and guidance for continuous monitoring, vulnerability management, and incident response.

- **National Risk Management Center (NRMC)**: Established within the CISA, the NRMC focuses on understanding and addressing the evolving risks to CI, including cybersecurity threats.

These elements highlight the comprehensive and collaborative approach that the United States has taken to safeguard CI from cyber threats. National cybersecurity strategies for CI continue to evolve to address the dynamic nature of cyber threats, enhance resilience, and ensure the continued availability and security of essential services.

Summary

In this chapter, we have gained insights into the transformative impact of intricate information infrastructures, including global computer networks and advanced control systems, on our modern era. These technologies have propelled our society to unprecedented levels of performance and have reshaped our culture. However, we have also learned about the need for vigilance due to the hidden vulnerabilities associated with our growing reliance on interconnected systems. The chapter has provided a glimpse into the history of cyberattacks on CI, the current global situation, and the national cybersecurity strategies.

In summary, this chapter has equipped us with a comprehensive understanding of the complex landscape of CI protection in an era of rapid technological advancement. It emphasizes the need for strategic planning, collaboration, and foresight to ensure the continued strength and security of our vital systems.

In the upcoming chapter, you can anticipate a thorough examination of security vulnerability assessment concepts. The chapter will provide a detailed exploration of the vulnerability life cycle and offer valuable insights into how to effectively assess and manage vulnerabilities.

References

To learn more about the topics that were covered in this chapter, take a look at the following resources:

- *National Commission on Terrorist Attacks Upon the United States. (2004). The 9/11 Commission Report*: https://www.9-11commission.gov/report/

- *Herodotus. (1998). The Histories. Oxford University Press.*

- *Sweet, K. M. (2009). Aviation and Airport Security: Terrorism and Safety Concerns. CRC Press.*

- *Zetter, K. (2014). Countdown to Zero Day: Stuxnet and the Launch of the World's First Digital Weapon. Crown.*

- *Lewis, J. A. (2014). Critical Infrastructure Protection in Homeland Security: Defending a Networked Nation. Wiley.*

- *The National Academies of Sciences, Engineering, and Medicine. (2017). Enhancing the Resilience of the Nation's Electricity Systems. The National Academies Press*: https://www.nap.edu/catalog/24836/enhancing-the-resilience-of-the-nations-electricity-system

- *Clinton, B. (1996). Executive Order 13010 – Critical Infrastructure Protection. Federal Register*: https://www.federalregister.gov/documents/1996/07/17/96-18351/critical-infrastructure-protection

- *Dunn Cavelty, M. (2008). Cyber-Security and Threat Politics: US Efforts to Secure the Information Age. Routledge.*

- *U.S. Department of Homeland Security. (2013). National Infrastructure Protection Plan*: https://www.cisa.gov/publication/nipp-2013-partnering-critical-infrastructure-security-and-resilience

- *The White House. (2018). National Cyber Strategy of the United States of America*: https://trumpwhitehouse.archives.gov/wp-content/uploads/2018/09/National-Cyber-Strategy.pdf

- *Perlroth, N. (2021). This Is How They Tell Me the World Ends: The Cyberweapons Arms Race.*

- *Kello, L. (2017). The Virtual Weapon and International Order. Yale University Press.*

- *Cybersecurity and Infrastructure Security Agency. (2023). CISA and Microsoft Partnership Expands Access to Logging Capabilities broadly*: https://www.cisa.gov/news-events/news/cisa-and-microsoft-partnership-expands-access-logging-capabilities-broadly

- *Federal Bureau of Investigation. (2018, November 2). The Morris Worm: 30 Years Since First Major Attack on the Internet*: https://www.fbi.gov/news/stories/morris-worm-30-years-since-first-major-attack-on-internet-110218

- *Lakshmanan, R. (2023, March). 3CX Supply Chain Attack: Here's What We Know*: https://thehackernews.com/2023/03/3cx-supply-chain-attack-heres-what-we.html

- *Lakshmanan, R. (2023, April). Lazarus Xtrader Hack Impacts Critical*: https://thehackernews.com/2023/04/lazarus-xtrader-hack-impacts-critical.html

- *National Cyber Security Centre. (2023, May). SVR cyber actors adapt tactics for initial cloud access*: https://www.ncsc.gov.uk/news/svr-cyber-actors-adapt-tactics-for-initial-cloud-access https://www.ncsc.gov.uk/news/svr-cyber-actors-adapt-tactics-for-initial-cloud-access

- *The White House. (2023). National Cybersecurity Strategy 2023*: https://www.whitehouse.gov/wp-content/uploads/2023/03/National-Cybersecurity-Strategy-2023.pdf

- *The White House, Office of the Press Secretary. (2013, February 12). Presidential Policy Directive - Critical Infrastructure Security and Resilience*: https://obamawhitehouse.archives.gov/the-press-office/2013/02/12/presidential-policy-directive-critical-infrastructure-security-and-resil

- *National Institute of Standards and Technology. (n.d.). Cybersecurity Framework*: https://www.nist.gov/cyberframework

- *Cybersecurity and Infrastructure Security Agency. (n.d.). C3 Voluntary Program for State, Local, Territorial, and Tribal Governments FAQ*: https://www.cisa.gov/sites/default/files/c3vp/sltt/CCubed_VP_FAQ.pdf

- *Cybersecurity and Infrastructure Security Agency. (n.d.). NCCIC Industrial Control Systems*: https://www.cisa.gov/sites/default/files/FactSheets/NCCIC%20ICS_FactSheet_NCCIC%20ICS_S508C.pdf

- *National Council of ISACs. (n.d.). Member ISACs*: https://www.nationalisacs.org/member-isacs-3

- *The White House. (2021, May 12). Executive Order on Improving the Nation's Cybersecurity*: https://www.whitehouse.gov/briefing-room/presidential-actions/2021/05/12/executive-order-on-improving-the-nations-cybersecurity/

- *Cybersecurity and Infrastructure Security Agency. (n.d.). Federal Government Cybersecurity Incident and Vulnerability Response Playbooks*: https://www.cisa.gov/resources-tools/resources/federal-government-cybersecurity-incident-and-vulnerability-response-playbooks

- *Cybersecurity and Infrastructure Security Agency. (n.d.). Enhanced Cybersecurity Services (ECS):* `https://www.dhs.gov/publication/dhsnppdpia-028a-enhanced-cybersecurity-services-ecs`

- *Cybersecurity and Infrastructure Security Agency. (n.d.). Continuous Diagnostics and Mitigation (CDM) Program:* `https://www.cisa.gov/resources-tools/programs/continuous-diagnostics-and-mitigation-cdm-program`

- *Cybersecurity and Infrastructure Security Agency. (n.d.). National Risk Management Center:* `https://www.cisa.gov/about/divisions-offices/national-risk-management-center`

- *Ottis, R. (2008). Analysis of the 2007 cyber attacks from the information warfare perspective. Cooperative Cyber Defence Centre of Excellence:* `https://www.ccdcoe.org/uploads/2018/10/Ottis2008_AnalysisOf2007FromTheInformationWarfarePerspective.pdf`

3

Critical Infrastructure Vulnerabilities

Welcome to this chapter, where we explore the world of security assessment concepts. In the realm of CI, understanding vulnerabilities and threats is of utmost importance. Throughout this chapter, we will explore the life cycle of a vulnerability and provide you with the skills to assess and manage it effectively.

Additionally, we will provide an overview of the most prevalent vulnerabilities and threats that are present in today's CI landscape, including industrial legacy infrastructure.

In this chapter, we will cover the following topics:

- Understanding the difference between threat, vulnerability, and risk
- Vulnerability assessment
- Most common vulnerabilities and threats in CI

By the end of this chapter, you will have a comprehensive understanding of the fundamentals of vulnerability assessment and be able to identify and address potential risks in critical systems.

Understanding the difference between threat, vulnerability, and risk

In the context of cybersecurity and risk management, the terms *vulnerability*, *threat*, and *risk* are often used interchangeably, but they represent distinct concepts. Understanding their differences is crucial for effectively managing and mitigating potential security issues. Let's explore each term with examples to illustrate their meanings.

Vulnerability

A **vulnerability** represents a weakness or deficiency within a system, application, or process that may be exploited by a threat to inflict harm. Vulnerabilities can be the result of design flaws, coding errors, misconfigurations, or outdated software. Identifying and fixing vulnerabilities is crucial to reducing the risk of security incidents.

Here are some examples:

Vulnerability	Description
Buffer overflow (coding error vulnerability)	An attacker submits a specially crafted input that exceeds the allocated buffer size, causing a buffer overflow. This vulnerability could allow the attacker to execute arbitrary code or gain unauthorized access to the system. For example, an application might accept user input without proper validation and sanitization.
Default credentials (misconfiguration vulnerability)	A network device is shipped with default login credentials (username and password) that are widely known. If administrators fail to change these defaults, an attacker can easily gain unauthorized access to the device and potentially compromise the entire network.
Unpatched software (outdated software vulnerability)	An organization's server runs an outdated version of a **content management system** (**CMS**) that contains known security vulnerabilities. Attackers exploit these vulnerabilities to gain control over the server, deface the website, or install malware, as the system hasn't received the necessary security patches.
SQL injection and XSS (design flaw vulnerabilities)	A web application fails to properly validate and sanitize user input, allowing for SQL injection or **cross-site scripting** (**XSS**) attacks. An attacker can insert malicious code into input fields, leading to database manipulation or the execution of malicious scripts in the context of other users' browsers.

Table 3.1 – Vulnerability examples

With a clearer understanding of the vulnerability concept, let us transition to an exploration of the subject of threats.

Threat

A **threat** encompasses any possible harmful occurrence that could exploit a vulnerability, leading to a compromise in the security of a system or organization.

Threats can be either intentional (malicious) or unintentional (accidental). They could come from external attackers, insiders, natural disasters, software bugs, or hardware failures. Here are some examples:

Threat	Potential Attack
A group of hackers	A group of hackers develop a sophisticated piece of malware and launch a cyberattack on a financial institution. The malware exploits a vulnerability in the bank's online banking system, compromising customer data and potentially leading to financial losses.
Employee mistake	An employee of a healthcare organization accidentally sends sensitive patient information, including medical records and personal details, to the wrong recipient due to a misconfigured email client. This unintentional action compromises patient privacy and violates data protection regulations.
Cybercriminals researching vulnerabilities	Cybercriminals discover a previously unknown software vulnerability (a zero-day) in a widely used operating system. They develop an exploit to take advantage of this vulnerability, gaining unauthorized access to systems, stealing sensitive information, and potentially causing widespread system breaches.
APT (Advanced Persistent Threat)	This can be any sophisticated and targeted cyberattack in which an unauthorized party gains access to a network or system with the intent to remain undetected for an extended period. APT attacks are characterized by their persistence, stealthiness, and often their association with nation-state actors, cyber espionage, or highly organized criminal groups.

Table 3.2 – Threats and potential attacks

Now that we have delineated the nature of threats, let us proceed to elucidate the concept of risk.

Risk

A **risk** is the potential harm or adverse consequence arising from the exploitation of a vulnerability by a threat. This concept combines the likelihood of a threat exploiting a vulnerability and the potential consequences if that were to occur:

$$Risk = Threat \ x \ Vulnerability = (low, moderate, high)$$

Risk assessment helps organizations prioritize security efforts and allocate resources effectively. Here are some hypothetical examples:

Asset	Risk	Likelihood
Seismic Monitoring Network	A seismic monitoring network is used to detect earthquake activity in a seismic-prone region. While there's a vulnerability in the network's software, the low likelihood of a cyber threat exploiting it due to the network's isolation from the internet and its strong access controls significantly reduces the risk.	Low
Airport Air Traffic Control	An airport's air traffic control system can rely on legacy software with known vulnerabilities. These vulnerabilities could be exploited by threat actors. The moderate likelihood of attacks targeting aviation infrastructure, combined with the potential consequences of disrupting air traffic, makes this risk moderate.	Moderate
Natural Gas Distribution System	A natural gas distribution system is monitored and controlled by industrial control systems. Vulnerabilities in these systems could be exploited, leading to gas leaks or pipeline disruptions. The high likelihood of cyberattacks targeting critical energy infrastructure increases the risk substantially.	High
National Power Grid	The national power grid's control systems have outdated and unsupported components. Known vulnerabilities in these components could be exploited by skilled threat actors. The very high likelihood of cyberattacks on a high-profile target with widespread consequences elevates the risk level significantly.	High
Water Supply Reservoir	A water supply reservoir's control systems are connected to the internet without proper security measures. Vulnerabilities exist in the control systems' architecture, and the extreme likelihood of attacks targeting critical water supply infrastructure puts the risk at an extreme level.	High

Table 3.3 – Examples of risks based on threats and vulnerabilities

These examples assess the likelihood of risk scenarios specific to CI systems, considering factors such as isolation from the internet, systems architecture, attractiveness to threat actors, and potential consequences of exploitation. The risk assessment helps organizations in the CI sector prioritize security efforts and allocate resources effectively to safeguard their systems and operations, with the ultimate priority being human safety.

It is also noteworthy to mention that risk assessments vary based on several factors, including the technical environment, the sector of the organization, and the historical context at the given time.

In summary, a *vulnerability* is a weakness that could be exploited by a threat. A *threat* is a potential malicious actor that could exploit a vulnerability. A *risk* is the measure of potential harm likelihood resulting from the interaction of a threat exploiting a vulnerability.

Vulnerability = Threat

Threat = Malicious actor

Risk = Likelihood of a threat exploiting a vulnerability

Understanding these distinctions is fundamental in developing a proactive and robust cybersecurity strategy. By identifying vulnerabilities, assessing associated risks, and implementing appropriate mitigation and/or remediation measures, organizations can reduce the likelihood and impact of security incidents. Regular vulnerability assessments and risk assessments that take relevant threats into account play a vital role in maintaining a strong security posture.

Having established clearer definitions of the terms threat, vulnerability, and risk, let us now explore the intricacies of the vulnerability assessment process.

Vulnerability assessment

The basics of vulnerability assessment involve a systematic and comprehensive approach to identifying, quantifying, and addressing vulnerabilities within a system, network, or organization. The primary goal of vulnerability assessment is to uncover weaknesses or security gaps that malicious actors could exploit to compromise the confidentiality, integrity, or availability of critical assets. Vulnerability assessment is a crucial process in cybersecurity that helps organizations identify and address potential weaknesses in their systems and networks. The following diagram shows the key steps and elements involved in vulnerability assessment:

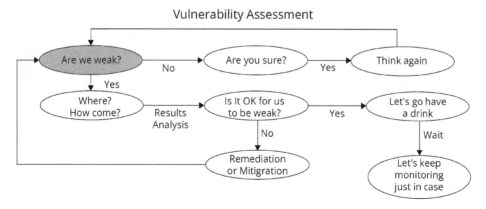

Figure 3.1 – Vulnerability assessment process

Here are the basics of vulnerability assessment, along with some examples to illustrate each step.

Scope definition

In this step, you define the scope and objectives of the assessment. Determine which systems, applications, or networks will be assessed, and what goals you want to achieve with the assessment. For example, you might decide to assess the security of your company's internal network, web applications, or cloud infrastructure.

Asset inventory

Create an inventory of all the assets within the defined scope. This includes hardware, software, devices, and critical data. For instance, you might list all the servers, routers, databases, web applications, and user endpoints that are relevant to the assessment.

Threat modeling

Conduct a threat modeling exercise to identify potential threats and attack vectors that could exploit vulnerabilities. This step helps you understand the risks your assets are exposed to. For example, you might identify threats such as unauthorized access attempts, SQL injection attacks, or phishing attempts against your users.

Vulnerability scanning

Use automated vulnerability scanning tools to scan the identified assets for known vulnerabilities. These tools check for missing patches, misconfigurations, and weak security settings. For example, a vulnerability scanner might find that a certain server is missing critical security updates or that a web application has weak password policies.

Manual assessment

Perform manual assessments to identify complex or unknown vulnerabilities that automated scanners might miss. Manual assessments involve in-depth analysis and testing by skilled security professionals. For instance, a manual assessment might include penetration testing, where ethical hackers attempt to exploit vulnerabilities in a controlled environment:

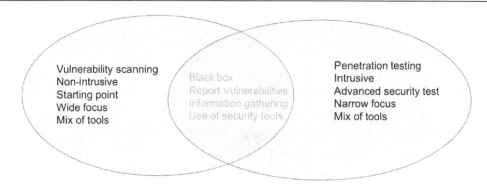

Figure 3.2 – Vulnerability scanning supported by penetration testing

Risk prioritization

After identifying vulnerabilities, prioritize them based on their severity and potential impact on your organization. High-risk vulnerabilities should be addressed urgently to minimize the chances of exploitation. For example, a critical vulnerability in a key database server should be given higher priority than a low-risk configuration issue.

Remediation planning

Develop a plan to address and mitigate the identified vulnerabilities. The plan should outline the necessary steps, the resources required, and a timeline for implementing fixes or countermeasures. For example, the plan might include applying security patches, updating software, or reconfiguring network settings.

Verification and validation

After implementing the remediation plan, verify that the vulnerabilities have been effectively mitigated. Reassess the systems to ensure the fixes were successful and no new vulnerabilities were introduced. For instance, you might run another vulnerability scan or perform additional penetration testing to confirm the effectiveness of the fixes.

Ongoing monitoring

Vulnerability assessment is not a one-time event. Establish a process for continuous monitoring and periodic reassessment to address new vulnerabilities that may arise due to system changes, updates, or emerging threats. Regular vulnerability scanning and security audits are examples of ongoing monitoring practices.

Reporting and documentation

Document all findings, assessment methodologies, and remediation plans. This documentation is essential for reporting to stakeholders, for compliance requirements, and as a reference for future assessments. The report should include a summary of vulnerabilities, their impact, and the actions taken to mitigate them.

By following these basics of vulnerability assessment and continuously improving the security posture, organizations can proactively protect their assets and data from potential cyber threats.

Let's see next what the vulnerability management life cycle is.

Security vulnerability management life cycle

The security vulnerability life cycle outlines the stages that a security vulnerability goes through from its discovery to its eventual resolution. Understanding this life cycle is essential for organizations to effectively manage and respond to security vulnerabilities. The typical vulnerability life cycle consists of the following stages:

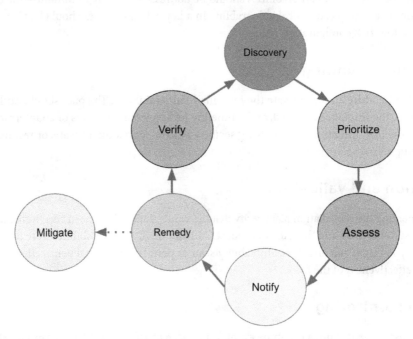

Figure 3.3 – Vulnerability management life cycle

Let's look at each of these stages.

Discovery

In this initial stage, a vulnerability is identified, either by security researchers, internal security teams, or even malicious actors. Vulnerabilities can be discovered through various means, such as security audits, vulnerability scanning, penetration testing, or incident investigations.

Assessment and prioritization

Once a vulnerability is discovered, it needs to be assessed to determine its severity and potential impact on the organization. Security teams and experts analyze the vulnerability to understand how it could be exploited and what assets or data are at risk. Vulnerabilities are then prioritized based on their criticality and potential impact on the organization's security.

Notification

After the vulnerability has been assessed and prioritized, affected vendors, developers, or system owners are notified. Responsible disclosure is typically followed, where the affected parties are informed about the vulnerability privately to give them time to develop and release a security patch or mitigation.

Remediation or mitigation

Users and administrators must take action to apply the security patch or mitigation to their systems. Timely remediation is crucial to protect the organization's assets from potential exploitation. Organizations may also implement temporary workarounds if a patch is not immediately available.

Verification and validation

After applying the security patch or mitigation, it is essential to verify its effectiveness and validate that the vulnerability has been successfully addressed. This verification process ensures that the organization is adequately protected against the specific vulnerability.

Monitoring and continuous assessment

Even after a vulnerability is patched or mitigated, the security team must continue to monitor the system for any signs of exploitation or new vulnerabilities that may arise due to system changes or updates. Regular security assessments and monitoring are critical for maintaining a strong security posture.

End of life

As software or systems age, vendors may eventually discontinue support or updates. When this happens, vulnerabilities may remain unpatched, leading to a situation where the vulnerability is considered critical and of the highest risk. In such cases, organizations must consider other measures, such as system upgrades or replacements, to mitigate risks effectively.

By understanding the security vulnerability life cycle, organizations can respond promptly to newly discovered vulnerabilities, implement necessary security measures, and reduce their exposure to potential security breaches. Proactive vulnerability management is crucial for maintaining the security and integrity of systems and data.

Most common vulnerabilities and threats in CI

CI is a prime target for cyber threats due to its significance and potential impact on society and the economy. In this section, we are going to cover some of the most common threats and vulnerabilities faced by CI, along with examples to illustrate each.

Inadequately secured industrial control systems (ICS)

Cyberattacks on ICS are malicious activities that target the computerized systems used to manage and control various industrial processes, such as manufacturing, energy production, water treatment, and more. These attacks aim to disrupt, manipulate, or gain unauthorized access to CI, posing significant risks to public safety, national security, and economic stability.

ICS plays a crucial role in automating and optimizing complex industrial operations, ensuring efficiency, accuracy, and safety in critical processes. ICS components include hardware, software, and networks that work together to monitor, control, and manage industrial processes.

There are several types of industrial control systems, each tailored to specific industrial sectors and processes. Here are some common types.

Supervisory control and data acquisition (SCADA) systems

SCADA systems are used to monitor and control industrial processes and infrastructure remotely. They collect real-time data from sensors, analyze it, and provide operators with a visual interface to monitor and control various processes. SCADA systems are commonly used in energy distribution, water treatment, and transportation systems.

Distributed control systems (DCS)

DCS are designed for complex and distributed industrial processes. They consist of a network of control units that collaborate to manage various parts of a process. DCS systems are used in industries such as chemical manufacturing, power generation, and oil refining.

Programmable logic controllers (PLCs)

PLCs are ruggedized computers that control specific processes or machinery. They are used to automate tasks by executing pre-programmed logic. PLCs are common in manufacturing industries and are essential for tasks such as robotics control, assembly line automation, and machinery operation.

Safety instrumented systems (SIS)

SIS are specialized systems that focus on ensuring the safety of industrial processes by taking predefined actions to prevent accidents or mitigate their consequences. These systems use sensors, logic solvers, and final control elements to manage critical safety processes.

Industrial Internet of Things (IIoT) systems

IIoT systems involve connecting various industrial devices, sensors, and machinery to the internet to collect and analyze data in real time. This data is used to optimize processes, predict maintenance needs, and improve overall operational efficiency.

Human-machine interface (HMI) systems

HMI systems provide a visual interface for operators to interact with and monitor industrial processes. They present real-time data and allow operators to control processes by sending commands to the underlying control systems.

Manufacturing execution systems (MES)

MES bridge the gap between enterprise-level systems and control systems. They manage production schedules, track material usage, monitor quality, and provide real-time visibility into manufacturing operations.

Building automation systems (BAS)

BAS are employed for the management and supervision of building systems, including heating, ventilation, air conditioning (HVAC), lighting, and security. They enhance energy efficiency and occupant comfort in commercial and industrial buildings.

Energy management systems (EMS)

EMS are employed in energy production and distribution to optimize power generation, transmission, and distribution. They help balance energy supply and demand, ensuring efficient energy utilization.

These various types of industrial control systems are designed to suit the specific needs of different industries, processes, and operational requirements, contributing to enhanced efficiency, safety, and productivity in industrial operations.

Common vulnerabilities in industrial control systems (ICS)

ICS form the backbone of numerous critical industries, and their vulnerabilities can lead to severe consequences. The following table focuses on dissecting the technical aspects of the most common vulnerabilities found in ICS environments:

Vulnerability	Description	Impact	Mitigation
Weak authentication and authorization	Many ICS components use default or weak credentials, and access control mechanisms are often lacking.	Unauthorized access can lead to the compromise of critical systems, disruption of processes, and unauthorized manipulation.	Implement strong authentication mechanisms, enforce access controls, and use **multi-factor authentication** (MFA) where possible.
Lack of patch management	ICS environments frequently use legacy systems that are not regularly updated, leaving them vulnerable to known exploits.	Unpatched systems are susceptible to attacks that exploit known vulnerabilities.	Establish a robust patch management process, prioritize critical updates, and consider virtual patching for legacy systems. If this is not possible then network isolate.
Inadequate network segmentation	Many ICS networks lack proper segmentation, allowing lateral movement of threats across different zones.	A successful breach in one part of the network can easily spread to other critical systems.	Implement a well-defined network segmentation strategy to limit communication between different parts of the network.
Vulnerable remote access points	Remote access points, such as VPNs, are often susceptible to known vulnerabilities.	Compromised remote access can provide attackers a foothold in critical systems and networks.	Regularly update remote access solutions, use strong authentication, and employ intrusion detection for remote access points.
Insufficient security monitoring	Inadequate monitoring of ICS networks and systems leads to delayed detection of anomalies and malicious activities.	Attackers can dwell within the network for extended periods, causing substantial damage before being detected.	Deploy **intrusion detection systems** (IDS) and **security information and event management** (SIEM) solutions to monitor and analyze network activities.

Vulnerability	Description	Impact	Mitigation
Lack of vendor security assessments	ICS components are often procured from third-party vendors without proper security assessments.	Vulnerable components introduced through the supply chain can serve as entry points for attackers.	Conduct thorough security assessments of vendor products, enforce security requirements in procurement contracts, and ensure timely updates from vendors.
Human error and social engineering	Inadequate security awareness and training can lead to human errors and susceptibility to social engineering attacks.	Attackers can exploit employees to gain unauthorized access or manipulate systems.	Implement regular security training for personnel, emphasize the importance of vigilant behavior, and conduct simulated phishing exercises.

Table 3.4 – Most common vulnerabilities in ICS environments

Understanding the technical intricacies of common vulnerabilities in ICS is paramount for effective risk mitigation. By addressing these vulnerabilities through robust cybersecurity practices, including strong authentication, proper network segmentation, timely patching, and comprehensive security assessments, industries can significantly enhance the resilience of their CI against cyber threats.

Ransomware targeting CI

Ransomware attacks have evolved into a significant threat vector, particularly when targeting CI sectors such as energy, transportation, and healthcare. The following table shows a typical ransomware attack flow:

Actions	Description	Impact	Mitigation
Initial infection vector	Attackers commonly use phishing emails, malicious attachments, or compromised websites to deliver ransomware payloads.	Unsuspecting users who interact with malicious content initiate the infection process.	Employ email filtering, educate employees about phishing, and update browsers and software regularly.

Actions	Description	Impact	Mitigation
Propagation and lateral movement	Once inside the network, ransomware seeks out vulnerable systems and spreads laterally to maximize impact.	Rapid lateral movement can lead to widespread encryption of critical systems.	Implement network segmentation, restrict lateral movement, and use endpoint detection and response (EDR) solutions.
Encryption of data	Ransomware encrypts files and data using strong encryption algorithms.	Encrypted data becomes inaccessible, disrupting operations and potentially causing downtime.	Regularly back up critical data offline, follow the 3-2-1 backup rule, and implement data loss prevention (DLP) solutions.
Ransom note and payment	After encryption, the attacker presents a ransom note with instructions for payment to decrypt files.	Paying the ransom may not guarantee successful decryption, and it supports criminal activities.	Prepare for potential attacks with a well-defined incident response plan and avoid paying ransoms.
Double extortion	Attackers threaten to leak stolen data if the ransom is not paid, increasing pressure on victims to comply.	Organizations may face public embarrassment, legal consequences, and regulatory fines.	Regularly update and patch systems, implement strong authentication, and segment networks.

Table 3.5 – Ransomware attack flow

Now, let's see another common attack on CI.

Supply chain attacks on CI components

Attackers may exploit vulnerabilities in the supply chain of CI components, introducing malicious code or compromising the integrity of the equipment or software.

Supply chain attacks targeting CI components have gained prominence as a potent threat vector. The following table explores the technical details of these attacks, outlining potential consequences and suggesting countermeasures to enhance supply chain security:

Attack	Description	Impact	Mitigation
Insertion of malicious code	Attackers compromise the supply chain by introducing malicious code into hardware or software components during development, manufacturing, or distribution.	Malicious code can compromise the functionality, integrity, and security of CI systems.	Implement code review processes, source code verification, and secure development practices.
Compromised software updates	Attackers target software update mechanisms to deliver compromised updates to CI components.	Compromised updates can introduce backdoors, vulnerabilities, or malware into critical systems.	Digitally sign software updates, use secure update channels, and verify software authenticity before deployment. Assess and audit the cybersecurity practices of third-party vendors and suppliers.
Hardware tampering	Attackers modify hardware components during the manufacturing process to include hidden vulnerabilities or backdoors.	Tampered hardware can compromise the security and integrity of CI systems.	Implement hardware integrity checks, conduct physical inspections, and source components from trusted suppliers. Implement hardware-based security features, including secure boot, **Trusted Platform Module** (TPM), and hardware-based attestation.
Insider threats within the supply chain	Insiders with access to the supply chain can deliberately introduce vulnerabilities or malicious components.	Insider actions can lead to the widespread compromise of CI systems.	Implement MFA for supply chain access and enforce strict access controls, conduct background checks on suppliers, and monitor supply chain activities.

Table 3.6 – Technical details of supply chain attacks

Supply chain attacks targeting CI components pose significant risks. Robust cybersecurity practices across the supply chain, including secure development, thorough vendor assessments, and hardware integrity checks, are vital to mitigating these threats. A proactive approach is essential to fortify the supply chain against vulnerabilities and maintain the security and integrity of CI components.

Legacy systems and lack of security updates

CI systems often use legacy equipment and software that may lack modern security features and updates, making them vulnerable to attacks. The following table details some vulnerabilities that arise as a consequence of inadequate patching:

Security Issue	Description	Impact	Mitigation
Outdated software and vulnerabilities	Legacy systems typically run outdated operating systems and software, often with known vulnerabilities that are no longer patched.	Attackers can exploit known vulnerabilities to compromise CI components and gain unauthorized access.	Migrate to modern systems, isolate legacy systems, and implement compensating security controls.
Unsupported hardware	Legacy systems may rely on hardware components that are no longer supported by manufacturers, making it difficult to address security vulnerabilities.	Insecure hardware can be exploited to compromise critical systems and introduce potential points of failure.	Consider hardware upgrades or replacements, and monitor security alerts for unsupported components.
Lack of security updates	Failure to apply security updates leaves systems exposed to known exploits and vulnerabilities.	Attackers can leverage unpatched vulnerabilities to compromise CI components.	Implement a robust patch management process, prioritize critical updates, and apply virtual patching if feasible.
Delayed response to threats	Legacy systems may have slower response times to emerging threats due to limitations in monitoring and detection capabilities.	Attacks can go undetected for longer periods, allowing attackers to maintain a foothold and cause damage.	Enhance monitoring capabilities, implement intrusion detection systems (IDS), and conduct regular security assessments.

Table 3.7 – Legacy system challenges and lack of security updates

Legacy systems and the lack of security updates pose significant risks to CI components. A proactive approach involving system modernization, regular updates, enhanced monitoring, and effective patch management is crucial to mitigating these vulnerabilities. By prioritizing security and investing in modern technology, CI sectors can bolster their defenses against evolving cyber threats.

Physical security breaches

Physical security breaches, such as unauthorized access to CI facilities, can lead to tampering with equipment or control systems, resulting in operational disruptions. The following table describes vulnerabilities attributable to the absence of adequate physical security measures:

Vulnerability	Description	Impact	Mitigation
Unauthorized access	Unauthorized individuals could gain access to CI facilities, equipment, or systems.	Intruders can manipulate systems, steal sensitive data, or cause disruptions.	Implement access controls, biometric authentication, and surveillance systems. Install barriers such as fences, bollards, and turnstiles to prevent unauthorized access.
Tailgating and piggybacking	Unauthorized personnel could follow authorized individuals into secure areas without proper authorization.	Intruders can gain entry to restricted zones and compromise critical systems.	Educate employees, deploy security personnel, and use turnstiles and access gates.
Insider threats	Employees with malicious intent or inadequate security awareness could exploit their access to perpetrate attacks.	Insider actions can lead to sabotage, unauthorized data access, or disruption of critical systems.	Implement role-based access controls, monitor employee activities, and conduct security awareness training.
Theft or damage of equipment	Attackers could steal or damage CI equipment, affecting operations.	Stolen or damaged equipment disrupts services and causes financial losses.	Implement asset tracking, secure storage, and CCTV surveillance.

Table 3.8 – Physical security breaches

Physical security breaches in CI can lead to severe consequences. Addressing vulnerabilities through a combination of access controls, surveillance, employee training, and incident response planning is vital to mitigating these risks and safeguarding critical systems against physical threats.

Internet of Things (IoT) vulnerabilities

The proliferation of IoT devices in CI introduces new attack surfaces, and insecurely configured or unpatched IoT devices can be exploited by attackers. The following table furnishes a description of the most prevalent vulnerabilities associated with IoT devices:

Vulnerability	Description	Impact	Mitigation
Insecure IoT device firmware	Many IoT devices ship with default or weak credentials, outdated firmware, and known vulnerabilities.	Attackers can exploit these vulnerabilities to gain unauthorized access and control over CI systems.	Regularly update firmware, change default credentials, and implement secure boot mechanisms.
Lack of encryption and data privacy	Inadequate data encryption during transmission and storage can expose sensitive information.	Unencrypted data can be intercepted, leading to data breaches and privacy violations.	Implement end-to-end encryption, secure data storage, and secure communication protocols.
Inadequate authentication and authorization	Weak authentication mechanisms and inadequate access controls can allow unauthorized access to IoT devices.	Unauthorized control of IoT devices can disrupt operations and compromise critical systems.	Implement strong authentication, role-based access controls, and MFA.
Lack of Over-the-Air (OTA) update security	IoT devices that lack secure OTA update mechanisms can be compromised through malicious updates.	Attackers can introduce malicious updates to compromise device functionality or exploit vulnerabilities.	Implement secure OTA update processes, code signing, and integrity checks.
Default services	IoT devices often have unnecessary services enabled by default, introducing potential attack surfaces.	Attackers can exploit these services to gain unauthorized access or launch attacks.	Disable unnecessary services, use device hardening techniques, and conduct regular vulnerability assessments.

Table 3.9 – IoT vulnerabilities

In summary, addressing IoT vulnerabilities demands a comprehensive strategy, including robust authentication, encryption, and vigilant patch management. As the IoT landscape evolves, proactive measures are crucial to ensure the ongoing security and resilience of interconnected systems. Adopting cybersecurity best practices remains pivotal to the secure integration of these technologies into our interconnected world.

These examples underscore the critical need for robust cybersecurity measures, continuous monitoring, and timely security updates within CI organizations. Collaborative efforts between the public and private sectors are also essential for sharing threat intelligence and best practices, ultimately strengthening the resilience of CI against emerging threats.

Summary

In this chapter on security vulnerability assessment concepts, we cover essential aspects of identifying and mitigating potential threats within CI. The chapter begins by explaining the distinction between threats and vulnerabilities, laying the groundwork for further exploration. The chapter emphasizes distinguishing between risk, vulnerability, and threat to foster a clear understanding of each element's significance in security assessments. Finally, we provide a comprehensive list and overview of the most common vulnerabilities and threats that exist within CI, with a specific focus on industrial legacy infrastructure. Armed with this knowledge, you will be better equipped to protect and manage critical systems effectively.

In the upcoming chapter, we explore the cybersecurity challenges faced by CI in today's interconnected world. Covering common cyber threats such as DDoS attacks and APTs, we dissect their mechanics and tactics, providing real-world examples. The next chapter aims to equip readers with a deep understanding of the evolving landscape of CI cybersecurity threats, offering insights into the complexities of defending vital infrastructure in our digitally intertwined age.

References

To learn more about the topics that were covered in this chapter, take a look at the following resources:

- *Cybersecurity and Infrastructure Security Agency. (n.d.). Defending Against Software Supply Chain Attacks*: https://www.cisa.gov/sites/default/files/publications/defending_against_software_supply_chain_attacks_508_1.pdf

- *National Cyber Security Centre. (n.d.). Understanding Vulnerabilities*: https://www.ncsc.gov.uk/collection/vulnerability-management/understanding-vulnerabilities#:~:text=A%20vulnerability%20is%20a%20weakness,to%20achieve%20their%20end%20goal

- *Department of Homeland Security. (2022). Ransomware Attacks*: https://www.dhs.gov/sites/default/files/2022-09/Ransomware%20Attacks%20.pdf

- *Cybersecurity & Infrastructure Security Agency. (2023). Cybersecurity Advisories [AA23-215A]:* `https://www.cisa.gov/news-events/cybersecurity-advisories/aa23-215a`

- *Cybersecurity & Infrastructure Security Agency. (2022). Cybersecurity Advisories [AA22-117A]:* `https://www.cisa.gov/news-events/cybersecurity-advisories/aa22-117a`

- *Cybersecurity & Infrastructure Security Agency. (2021). Cybersecurity Advisories [AA21-209A]:* `https://www.cisa.gov/news-events/cybersecurity-advisories/aa21-209a`

- *Tal, J. (September 2018). America's Critical Infrastructure: Threats, Vulnerabilities, and Solutions. Security InfoWatch:* `https://www.securityinfowatch.com/access-identity/access-control/article/12427447/americas-critical-infrastructure-threats-vulnerabilities-and-solutionshttps://www.securityinfowatch.com/access-identity/access-control/article/12427447/americas-critical-infrastructure-threats-vulnerabilities-and-solutions`

- *Labus, H., (2022, March 15). Critical Infrastructure Security. Help Net Security:* `https://www.helpnetsecurity.com/2022/03/15/critical-infrastructure-security/`

- *Wang, B., (2022, February 14). Types of Cyber Vulnerabilities. CrowdStrike:* `https://www.crowdstrike.com/cybersecurity-101/types-of-cyber-vulnerabilities/`

- *Federal Emergency Management Agency. (n.d.). National Response Framework:* `https://www.fema.gov/emergency-managers/national-preparedness/frameworks/response`

- *Department of Homeland Security. (2011). Risk Management Fundamentals: Homeland Security Risk Management Doctrine:* `https://www.dhs.gov/xlibrary/assets/rma-risk-management-fundamentals.pdf`

- *U.S. Government Accountability Office. (2023). CI Protection: National Cybersecurity Strategy Needs to Address Information Sharing Performance Measures and Methods:* `https://www.gao.gov/products/gao-23-105468` `https://www.gao.gov/products/gao-23-105468`

- *Haughey, C.J., Security Intelligence. (2023). The growing threat to CI:* `https://securityintelligence.com/articles/growing-threat-to-critical-infrastructure/` `https://securityintelligence.com/articles/growing-threat-to-critical-infrastructure/`

- *Carnegie Mellon University. (2016). CRR Resource Guide - Vulnerability Management*: `https://www.cisa.gov/sites/default/files/publications/CRR_Resource_Guide-VM_0.pdf https://www.cisa.gov/sites/default/files/publications/CRR_Resource_Guide-VM_0.pdf`

- *European Union Agency for Cybersecurity (ENISA). (n.d.). Vulnerabilities and Exploits*: `https://www.enisa.europa.eu/topics/incident-response/glossary/vulnerabilities-and-exploits`

Part 2:
Dissecting
Cyberattacks on CI

Part 2 is an in-depth exploration of cyber threats to critical infrastructures, detailing the methodologies of common attacks and profiling the attackers. It expands your understanding by examining case studies of significant cyber incidents, providing a practical framework for applying cybersecurity strategies to real-world scenarios. This section is crucial for those seeking to deepen their technical knowledge and enhance their defenses against the myriad cyber threats facing critical infrastructure.

This part has the following chapters:

- *Chapter 4, The Most Common Attacks Against CI*
- *Chapter 5, Analysis of the Top Cyber-Attacks on Critical Infrastructure*

4

The Most Common Attacks Against CI

In an increasingly interconnected and digitized world, **critical infrastructure (CI)** stands at the nexus of modern society's functionality. Comprising sectors such as energy, transportation, water supply, and healthcare, CI plays a pivotal role in sustaining our daily lives. However, this dependence on technology and interconnected systems also exposes CI to a growing threat landscape in the realm of cybersecurity. As nations and organizations grapple with the challenges of safeguarding these vital assets, the need to understand and combat cyber threats has never been more critical.

This chapter embarks on a comprehensive exploration of the most common cyberattacks that pose a significant risk to CI worldwide. From **distributed denial of service (DDoS)** assaults that flood networks with malicious traffic to **advanced persistent threats (APTs)** that stealthily infiltrate and compromise systems, each of these cyberattacks is dissected and examined. We will dig into the intricate mechanics of these threats, shedding light on how they activate, operate, and ultimately succeed in their nefarious objectives.

By providing real-world examples and dissecting the tactics, techniques, and procedures employed by threat actors, this chapter aims to equip readers with a profound understanding of the ever-evolving landscape of CI cybersecurity threats. As we journey through the various attack vectors, readers will gain valuable insights into the complexities of defending CI in an age where the digital realm and the physical world are increasingly intertwined.

In this chapter, we will cover the following topics:

- DDoS attack
- Ransomware attacks
- Supply chain attacks
- APTs

- Phishing
- Common unpatched vulnerability exploits

DDoS attack

A DDoS attack maliciously aims to interrupt the normal operations of a network, service, website, or online platform by inundating it with excessive traffic from numerous sources. The goal of a DDoS attack is to render the targeted system or service unavailable to its intended users temporarily or, in some cases, for an extended period. Here's how it works in two basic steps:

- **Step 1 – Multiple attack sources**: During a DDoS attack, the perpetrator generally employs a botnet, a network of hijacked computers, to create an overwhelming amount of traffic.
- **Step 2 – Traffic overload**: A network of compromised computers sends a flood of requests or data packets, sometimes malformed, to the target simultaneously. The target's servers or infrastructure become overwhelmed, causing a slowdown or complete disruption of services.

The following diagram illustrates a typical DDoS attack:

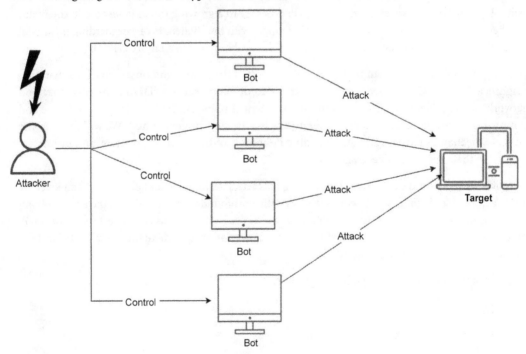

Figure 4.1 – DDoS attack

DDoS attacks come in various types, each with its own techniques and characteristics. The following are some technical types under which DDoS attacks can be classified.

Volumetric attacks

DDoS attacks can be classified as **volumetric attacks**. Volumetric attacks are a type of DDoS attack that focuses on overwhelming a target system or network with an extremely high volume of traffic. These attacks aim to consume the available bandwidth or exhaust the target's computational resources, making it difficult or impossible for legitimate users to access the targeted service. Volumetric attacks are one of the most common types of DDoS attack and can be highly disruptive. Here are the key characteristics and subtypes of volumetric attacks:

Type	Description	How to
UDP flood	Attackers flood the target with a high volume of **User Datagram Protocol (UDP)** packets, overwhelming the network bandwidth. UDP is a connectionless protocol, which means it does not establish connections like TCP does. This lack of connection tracking makes UDP particularly susceptible to flooding attacks because there's no state information to manage or validate incoming packets.	Attackers can use reflection and amplification techniques to make UDP flood attacks even more potent. They send small UDP packets with a spoofed source IP address to vulnerable servers (for example, DNS, NTP, or Memcached servers). These servers then respond to the victim with larger UDP responses, amplifying the attack traffic.
TCP flood	These are like UDP floods, but target the **Transmission Control Protocol (TCP)** to exhaust server resources.	The most common type of TCP flood attack is the SYN flood. In a normal TCP connection setup, a client sends a SYN packet to the server, the server acknowledges it with a SYN-ACK packet, and the client responds with an ACK packet to complete the connection. The attacker instructs the compromised devices in the botnet to send many SYN packets with a spoofed IP address to the target server. The server allocates resources (such as memory and CPU) for each incoming SYN request, expecting the connection to complete.

Type	Description	How to
ICMP flood	Attackers send many **Internet Control Message Protocol (ICMP)** packets, often used in **ping attacks**, to saturate network resources.	ICMP is commonly used for network diagnostics, and a flood of ICMP packets can consume a target's network bandwidth and resources, rendering it unresponsive. The attacker can command compromised devices in a botnet to send many ICMP Echo Request packets (ping requests) to the target. These packets are typically sent at a high rate and in a continuous fashion. As the target device or network receives a flood of ICMP Echo Requests, it must process each request and generate a response (ICMP Echo Reply). This consumes network bandwidth and processing resources, potentially leading to network congestion and resource exhaustion.

Table 4.1 – Volumetric attacks

Let's look at reflection and amplification attacks now.

Reflection and amplification attacks

Here are the key characteristics and subtypes of reflection and amplification attacks:

Type	Description	How to
DNS amplification	Attackers exploit vulnerable DNS servers to amplify traffic and direct it toward the target, causing congestion.	An attacker finds open DNS resolvers, which are publicly accessible DNS servers that don't restrict incoming requests. These servers are then used to direct a DNS amplification attack. The attacker spoofs their DNS query packets to appear as if they're coming from the target's IP address and sends them to the open resolvers. Due to the nature of DNS responses being larger than queries, this causes an amplification effect. A small query generates a significantly larger response. As a result, the open DNS resolvers, deceived into believing the target is requesting information, flood the target's IP address with large DNS responses. This overwhelming flow of data can disrupt the target's network services.
NTP amplification	Attackers misuse **Network Time Protocol** (**NTP**) servers to amplify attack traffic.	The attacker identifies vulnerable NTP servers on the internet. These are NTP servers configured to respond to NTP queries from any source, often due to misconfiguration or lack of security controls. The attacker sends numerous NTP queries to the vulnerable server with a forged source IP address that matches the targets. This makes it appear as if the target is requesting NTP synchronization information. NTP servers, responding to these queries, generate NTP responses that are significantly larger than the queries, creating an amplification effect. This results in the targeted NTP servers sending large responses to the victim's IP address, overwhelming its network resources.

Type	Description	How to
SSDP/UPnP reflection	Attackers exploit vulnerable **Simple Service Discovery Protocol** (**SSDP**) and **Universal Plug and Play** (**UPnP**) devices to reflect traffic.	The attacker sends many SSDP/UPnP discovery requests to vulnerable internet devices with the source IP address in these requests spoofed to appear as if they're coming from the target. These requests are typically UDP packets sent to port 1900, the standard port used for SSDP. SSDP-enabled devices, upon receiving the discovery requests, respond with information about their services. These responses can be much larger in size than the original requests, depending on the complexity of the device and the services it offers. The responses generated by the vulnerable devices are much larger than the initial discovery requests. This makes SSDP-enabled devices excellent amplifiers for DDoS attacks.
Memcached amplification	Attackers abuse unprotected Memcached servers to amplify DDoS traffic.	The attacker spoofs the source IP address in their Memcached requests. The attacker sends many Memcached GET requests to the publicly accessible and not properly secured Memcached servers previously identified. These requests are typically UDP packets sent to port 11211, the default port used for Memcached. Memcached servers respond to GET requests with the requested data. However, Memcached responses can be significantly larger in size than the corresponding requests, especially when dealing with large data objects or multiple objects. This amplification effect is what makes Memcached amplification attacks potent. Open Memcached servers, mistaking the target for a data requester, send substantial responses to the target's IP, leading to network overload. This results in high bandwidth usage and server resource depletion, causing service disruption for legitimate users.

Table 4.2 – Reflection and amplification attacks

Next, let's explore resource depletion attacks.

Resource depletion attacks

Here are the key characteristics and subtypes of depletion attacks:

Type	Description	How to
Ping of death	Attackers send oversized ICMP packets that cause buffer overflows or crashes on the target system.	The attacker typically needs a tool or script capable of crafting oversized ICMP packets. The attacker crafts ICMP Echo Request packets (ping packets) with deliberately forged or oversized payloads. The payload is intentionally made larger than the **maximum transmission unit** (**MTU**), which is the maximum size a packet can be on a given network. This oversized payload can cause problems when the target device attempts to reassemble and process the packet. The attacker sends the malformed ICMP ping packets to the target's IP address. These packets are typically transmitted at a high rate to flood the target. When the target receives the oversized ICMP packets, its networking or operating system stack may not properly handle the packet's size. This can lead to various consequences, including the following: • **Buffer overflows**: The target's buffer for processing ICMP packets may overflow, causing memory corruption • **Resource exhaustion**: The target may consume excessive CPU and memory resources attempting to process the malformed packets • **Crashes**: In some cases, the target's networking stack or the entire operating system may crash or become unresponsive

Type	Description	How to
Fragmentation attacks	Attackers send fragmented packets that confuse network devices and require extra processing resources to reassemble.	In networking, data travels in packets, sized to match the MTUs of different network paths. MTUs limit the largest packet size for each network segment, preventing fragmentation.
		The attacker designs a packet to exploit network fragmentation, intentionally making it break into smaller pieces for malicious reassembly. This packet, comprising several fragments, is sent to the target system or network.
		The target system processes incoming fragmented packets, reassembling them into their original form. Attacks can occur based on reassembly: overlapping fragments by the attacker can cause system confusion and vulnerabilities, incomplete reassembly might lead to data misinterpretation or loss, and out-of-order fragments can burden the system with reordering, potentially leading to packet rejection.

Table 4.3 – Resource depletion attacks

Let's explore protocol-based attacks in the next section.

Protocol-based attacks

Protocol-based DDoS attacks are a category of DDoS attack that exploit vulnerabilities or limitations in the protocols used by target servers or network devices. These attacks aim to exhaust the target's resources or disrupt its ability to establish and maintain network connections effectively. Here are some common types of protocol-based DDoS attacks:

Type	Description	How to
HTTP/HTTPS flood	In this attack, attackers send a massive number of HTTP or HTTPS requests to a web server. These requests can be legitimate GET or POST requests but are sent at an extremely high rate, overwhelming the server's capacity to handle them and causing service disruption.	The attacker instructs the compromised devices in the botnet to send a high volume of HTTP or HTTPS requests to the target. These requests are typically crafted to resemble legitimate user requests, making it more challenging to distinguish them from real traffic. To make it harder to trace the attack back to the source and to bypass rate-limiting or blocking mechanisms, the attacker may use spoofed or randomized source IP addresses in the requests. As the target web server or application receives a flood of HTTP/HTTPS requests, it begins to allocate resources (such as CPU, memory, and network bandwidth) to handle each incoming request. The volume of requests can quickly overwhelm the server's capacity to respond to legitimate traffic.

Type	Description	How to
Slowloris attack	This attack is a form of HTTP flood that is designed to be stealthy. Attackers send partial HTTP requests and keep the connections open by sending occasional data, preventing the server from closing the connections. This gradually ties up server resources and can lead to a denial of service.	The attacker uses a script or tool (Slowloris) to initiate multiple HTTP connections to the target web server. Once the connections are established, the attacker sends partial HTTP requests to the target server. These partial requests contain valid HTTP headers, such as GET, POST, or HEAD, but the request is intentionally left incomplete. For example, the attacker may not send the final HTTP newline character (\r\n\r\n) that signals the end of a legitimate HTTP request.

After sending the partial request, the attacker's connections remain open and idle. Importantly, the attacker sends periodic small additional data to keep these connections alive. These small *header* packets may contain headers such as Host or User-Agent.

Slowloris keeps these connections open, gradually consuming available resources, such as connection slots or worker threads, without completing any HTTP requests. As more connections are opened by the attacker and kept in an idle state, the target server's performance degrades, eventually making the web service unavailable. |

Type	Description	How to
RST (Reset) attack	In a TCP RST attack, attackers send a high volume of forged TCP RST packets to terminate established connections abruptly. This can disrupt legitimate communications and result in service interruptions.	In typical network communication, two devices establish a TCP connection to exchange data. This involves a series of steps, including the initial three-way handshake (SYN, SYN-ACK, ACK), to set up the connection. The attacker identifies a TCP connection between two devices that they want to disrupt. They may do this through network scanning, monitoring, or other means.
		The attacker crafts malicious TCP **Reset (RST)** packets with the same source and destination IP addresses and port numbers as the legitimate devices in the established connection. These RST packets are designed to appear as if they were sent by one of the legitimate communication partners.
		The attacker sends the forged RST packets into the network, targeting the established connection. These RST packets try to trick one or both legitimate devices into prematurely terminating the connection. When one of the legitimate devices receives the forged RST packet, it interprets it as a signal to reset (terminate) the connection immediately. This results in the TCP connection being abruptly terminated, and any ongoing communication is disrupted.

Type	Description	How to
ACK attack	Attackers send many TCP ACK (acknowledge) packets to a target server. While ACK packets are typically part of the normal TCP handshake, an excessive volume of ACK packets can consume server resources, slowing down or disrupting network communication. In a legitimate TCP connection establishment, a three-way handshake occurs: • **Step 1 (SYN):** The client sends a TCP packet with the SYN (synchronize) flag set to the server, indicating its intention to initiate a connection. • **Step 2 (SYN-ACK):** The server responds with a TCP packet that has both the SYN and ACK (acknowledge) flags set, indicating it's willing to establish a connection. • **Step 3 (ACK):** The client responds with a TCP packet that has only the ACK flag set, indicating acknowledgment of the server's response. The connection is now established.	The attacker generates a massive volume of TCP ACK packets, typically without completing the full three-way handshake. These packets contain the ACK flag set but lack the SYN flag, which is necessary for a valid three-way handshake. The attacker floods the target server or network device with these malicious TCP ACK packets. The target may receive a large number of ACK packets without corresponding SYN packets, which is unusual in normal network traffic. The target device's TCP stack attempts to process each incoming ACK packet and tries to match it with a valid connection. This consumes processing power and memory resources, which may result in resource exhaustion.

Type	Description	How to
SSL/TLS renegotiation attack	Attackers exploit vulnerabilities in the SSL/TLS renegotiation process to overwhelm a target server's CPU resources, making it difficult for the server to process legitimate SSL/TLS handshake requests. In a typical SSL/TLS connection, the client and server perform an initial handshake to establish a secure encrypted connection. During this handshake, the encryption parameters, including the cryptographic keys and algorithms, are negotiated. After the initial handshake is complete, data can be securely exchanged between the client and server over the established encrypted connection. Sometimes, there is a need to renegotiate the SSL/TLS parameters during an active connection. This is typically done to update encryption settings or perform certain operations such as client certificate authentication. Renegotiation begins with a new handshake within the existing connection.	The attacker acts as a client and initiates an SSL/TLS renegotiation request to the server. This request is a legitimate part of the SSL/TLS protocol and is meant to allow the client and server to renegotiate encryption parameters. The attacker sends multiple renegotiation requests to the server, creating a situation where multiple renegotiations are pending simultaneously. Renegotiating SSL/TLS parameters is a computationally intensive process for the server, especially when multiple renegotiations are pending. The server can become overwhelmed by the high volume of renegotiation requests, leading to excessive CPU and memory usage.

Type	Description	How to
Out-of-state packets	Attackers send packets with incorrect or out-of-sequence TCP/IP header information. The target server expends resources attempting to process these malformed packets, which can lead to service degradation or disruption.	An attacker attempts to disrupt or compromise network communication by sending packets that do not conform to the expected state or context of an ongoing session. This can include the following: • Sending packets that do not belong to an established session • Sending packets with incorrect or invalid sequence numbers, which are used to maintain the order of TCP segments • Sending packets with unexpected flags, options, or content • Sending packets targeting closed or nonexistent sessions • Sending malformed or forged packets that are designed to trigger errors or vulnerabilities in the network security device

Table 4.4 – Protocol-based attacks

Let's explore application layer attacks next.

Application layer attacks

Here are the key characteristics and subtypes of application layer attacks:

Type	Description	How to
SQL injection	Attackers inject malicious SQL queries into input fields to disrupt database operations.	An attacker can inject malicious SQL statements into an application's input fields, often in search or filter functions, that are used to construct database queries. These injected queries may be designed to retrieve an excessive amount of data, perform complex calculations, or join multiple tables, consuming a significant amount of database resources. Maliciously crafted SQL queries can be slow to execute, especially if they involve operations such as sorting or filtering large datasets. As a result, legitimate database queries may be delayed or even time out, causing a DoS condition. Extremely malicious SQL injection attacks can lead to database crashes or errors if they exploit vulnerabilities in the database management system. A crashed database becomes temporarily unavailable to all users, leading to a DoS.

Type	Description	How to
XML-RPC and SOAP attacks	Attackers exploit vulnerable XML-based services to overload servers with XML-RPC or SOAP requests.	XML-RPC is a remote procedure call protocol encoded in XML. It allows a client to invoke methods on a server remotely by sending an XML request over HTTP or other transport protocols.
		SOAP is another protocol for exchanging structured information in web services. It uses XML to send requests and responses between client and server applications.
		Attackers may attempt to inject malicious data into XML-RPC or SOAP requests, exploiting vulnerabilities in the way the server processes XML data. This can lead to issues such as SQL injection or remote code execution if the server processes XML data unsafely. Attackers can flood XML-RPC or SOAP endpoints with a high volume of requests, overwhelming the server and causing a denial of service.

Type	Description	How to
Zero-day exploits	Attackers target vulnerabilities that are not yet known or patched by software vendors.	Attackers search for previously unknown vulnerabilities in software, operating systems, or hardware components. This can involve reverse engineering, fuzz testing, or other techniques to identify weak points in a system's security. Once a vulnerability is discovered, attackers develop an exploit for it. An exploit is a piece of code or a technique that takes advantage of the vulnerability to gain unauthorized access, execute malicious code, or perform other malicious actions. Attackers select their target, which could be a specific organization, industry, or software product. The choice of target depends on the potential value of the attack. The attacker launches the zero-day attack against the target. This could involve delivering a malicious payload through various means, such as email attachments, infected websites, or network attacks. If successful, the exploit allows the attacker to execute a payload on the target system. The payload can vary in functionality, from stealing sensitive data to establishing a backdoor for future access.

Type	Description	How to
Bypassing rate limiting	Attackers use sophisticated techniques to bypass rate limiting and anti-DDoS measures.	Attackers may use multiple accounts or IP addresses to distribute requests across a range of sources. By doing this, they can make it appear as though the requests are coming from different users, making it harder for rate-limiting mechanisms to detect the attack.
		Attackers may insert artificial delays between their requests to stay within the rate limit. This way, they can continue making requests over an extended period without exceeding the allowed rate.
		Attackers can employ proxy servers or anonymization services to hide their true IP addresses. By routing requests through multiple proxies, they can appear as if they are coming from different locations, making it challenging to enforce rate limits based on IP addresses.
		If rate limiting is based on user sessions or tokens, attackers may manipulate or forge session tokens to appear as different users or sessions. Some rate-limiting mechanisms rely on counters or timers.
		Attackers may exploit race conditions by making requests concurrently or in quick succession, causing the rate-limiting counter or timer to reset before it can enforce the limit.

Table 4.5 – Application layer attacks

These are just some examples of the many DDoS attack techniques that malicious actors employ. To defend against these attacks, organizations implement a combination of network security measures, traffic monitoring, intrusion detection systems, and traffic scrubbing services. Regularly updating software and hardware, as well as staying informed about emerging threats, is essential for effective DDoS mitigation.

For CI, such as power grids, financial systems, transportation networks, and emergency services, a DDoS attack can have potentially devastating consequences:

- **Service disruption**: DDoS attacks can disrupt essential services, causing power outages, financial system failures, transportation delays, or communication breakdowns. This can lead to significant economic losses and public inconvenience.

- **Loss of data and control**: Attackers may use DDoS attacks as a diversionary tactic to distract security personnel while they attempt to breach CI systems. During the chaos of an attack, they may gain unauthorized access to sensitive data or even take control of systems.

- **Public safety risks:** In some cases, DDoS attacks against CI can pose direct risks to public safety. For instance, if an attack affects emergency communication systems or disrupts traffic control systems, it could lead to accidents, injuries, or worse.

- **Long-term damage**: DDoS attacks can cause long-lasting reputational damage to organizations responsible for CI. It can erode public trust in the reliability and security of essential services.

In this section, we looked at different types of DDoS attacks that pose significant threats to digital infrastructure. Up next, we'll explore the complexities of ransomware attacks.

Ransomware attack

A ransomware attack is a type of malicious cyberattack in which cybercriminals encrypt a victim's data or computer systems and then demand a ransom (usually in cryptocurrency) in exchange for providing the decryption key or restoring access to the compromised systems. Ransomware attacks are financially motivated and can have severe consequences for individuals, businesses, and organizations. The following picture illustrates ransomware attack activities:

Figure 4.2 – Ransomware attacks (source: Freepik.com)

Here's how a typical ransomware attack unfolds.

Infection

Ransomware is typically spread through malicious email attachments, through malicious links in emails or websites, or by exploiting software vulnerabilities. For example, when a user opens an infected file or clicks on a malicious link, the ransomware payload is executed on the victim's computer.

Encryption

Once the ransomware is executed, it begins encrypting files on the victim's computer or network. This encryption process renders the victim's data inaccessible and unreadable without the decryption key, which is held by the attackers.

Ransom note

After encrypting the victim's files, the ransomware displays a ransom note on the victim's screen, informing them that their data has been encrypted and explaining the steps they need to take to pay the ransom and receive the decryption key.

Ransom payment

The attackers demand a ransom payment, usually in cryptocurrency (such as Bitcoin or Monero), in exchange for the decryption key. Payment instructions and a deadline for payment are provided in the ransom note.

Data recovery

Should the victim choose to comply with the demand and transfer the cryptocurrency to the provided wallet address, the attackers, upon verification of the payment, might then give the decryption key to the victim. This key can potentially enable the victim to regain access to their encrypted data.

No guarantee of data recovery

It is important to recognize that succumbing to ransom demands does not provide any guarantees. Paying the ransom does not ensure the recovery of your data or the delivery of a valid decryption key from the attackers. In numerous instances, victims who paid the ransom found themselves facing not only data loss but also financial loss, as the promised solutions were not delivered.

Moreover, paying a ransom directly supports and fuels the activities of cybercriminals. Ransomware attacks are, by nature, unlawful and malicious acts. When victims pay ransoms, they inadvertently contribute to the profitability of these criminal endeavors. This financial incentive encourages cybercriminals to continue their illegal activities, putting countless others at risk of falling victim to similar attacks.

Choosing not to pay the ransom can send a powerful message. By refusing to meet the demands, victims make ransomware attacks less lucrative for cybercriminals. This can serve as a deterrent, dissuading attackers from targeting other individuals or organizations.

Paying ransoms perpetuates a lack of accountability among cybercriminals. It essentially conveys the message that these criminals can continue their activities without fear of facing consequences for their actions.

It's also important to consider the long-term impact. Paying ransom not only supports the individual attackers but also attracts new criminals into the ransomware business. The prospect of quick and easy money is a strong motivator for cybercriminals, and it perpetuates the cycle of attacks.

Equally important, paying a ransom doesn't address the underlying security weaknesses that allowed the ransomware attack to occur in the first place. To prevent future attacks, organizations must focus on enhancing their overall cybersecurity posture, which includes patching vulnerabilities, implementing robust security measures, and educating employees on best practices.

The decision to pay a ransom can have legal and ethical implications. Some jurisdictions have laws against making payments to cybercriminals, and organizations may face reputational damage and legal consequences for participating in such activities. It's worth noting that ransom payments can sometimes indirectly support terrorism or other illicit activities. The path of ransom funds can be challenging to trace, potentially contributing to nefarious purposes.

In lieu of paying a ransom, victims are encouraged to explore alternative approaches. These may include reporting the attack to law enforcement agencies, restoring systems and data from secure backups, seeking professional assistance from cybersecurity experts, and investing in enhancing overall security measures. These actions align with principles of security, legality, and ethical responsibility, ultimately contributing to a safer digital environment for all.

To protect against ransomware attacks, individuals and organizations should take proactive cybersecurity measures, including regularly updating software, using strong and unique passwords, implementing robust cybersecurity solutions, and educating users about phishing and safe internet practices. Additionally, maintaining up-to-date backups of critical data is essential to facilitate recovery without paying a ransom.

Supply chain attack

Supply chain attacks are a type of cyberattack that targets an organization by exploiting vulnerabilities or weaknesses in its supply chain or trusted third-party partners. These attacks occur when malicious actors compromise a supplier, service provider, or partner organization to gain unauthorized access to the target organization's systems, data, or infrastructure. Supply chain attacks can have serious consequences and are a growing concern in cybersecurity.

The recent surge in supply chain attacks has raised significant concerns, particularly regarding CI. These attacks have grown in frequency and sophistication over the past few years, presenting a clear and present danger to organizations across various sectors.

The proliferation of supply chain attacks can be attributed to several interconnected factors. One major driver is the increasingly complex and interdependent nature of modern business ecosystems. Organizations now rely heavily on extensive networks of suppliers, vendors, and partners to deliver goods and services efficiently. This interconnectedness, while beneficial for business operations, has created a vast and intricate attack surface. Each link in the supply chain can potentially serve as an entry point for cybercriminals seeking unauthorized access.

The accelerating pace of digital transformation has expanded this vulnerability. As organizations adopt cloud-based services, digital platforms, and third-party technology solutions, they often rely on external providers for critical functions. Cyberattackers have seized upon this trend, recognizing that targeting supply chain partners can yield significant rewards by compromising systems, data, or infrastructure.

The complexity inherent in software development and distribution processes has also played a pivotal role. Collaborative software development often involves contributions from multiple teams or individuals located across diverse geographic locations. This complexity can introduce vulnerabilities or malicious code into the supply chain, providing cybercriminals with opportunities to exploit these weaknesses.

Cybercriminals have also grown more sophisticated in their tactics, leveraging advanced tools, social engineering techniques, and thorough reconnaissance efforts. Moreover, nation-state actors and APT groups have identified the strategic potential of supply chain attacks for purposes such as espionage, surveillance, or infrastructure disruption, further escalating the threat landscape.

High-profile supply chain attacks have brought the issue to the forefront of cybersecurity concerns. Notable incidents such as the SolarWinds breach have underscored the extensive and far-reaching consequences of a compromised supply chain. These attacks have raised awareness of the profound impact and severity of supply chain vulnerabilities for organizations and their stakeholders.

The economic motivation of cybercriminals has also been a driving force behind these attacks. Attackers are increasingly drawn to supply chain compromises as a means of financial gain. They seek to profit from data theft, ransom demands, or manipulation of financial transactions, making supply chain attacks a lucrative option.

Amid these concerns, CI has become a particularly vulnerable target. The reliance on essential services, such as power grids, water treatment facilities, transportation networks, and emergency services, on interconnected and digitized systems makes them attractive targets for cyber adversaries. A successful supply chain attack on CI can have dire consequences, including service outages, public safety risks, and economic disruption. The following picture illustrates how an attacker typically performs a supply chain attack:

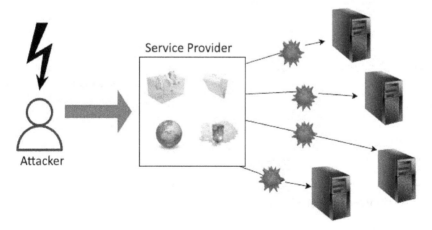

Figure 4.3 – Supply chain attack

The main characteristics of supply chain attacks are as follows.

Scope of attack

Supply chain attacks can affect a wide range of organizations, from small businesses to large enterprises. The attack surface extends beyond the target organization to include suppliers, vendors, contractors, and other third-party entities that have access to the organization's systems or data.

Attack vector

Malicious actors often exploit vulnerabilities in the software, hardware, or network infrastructure provided by a third-party supplier or partner. Common vectors include tainted software updates, compromised hardware components, or maliciously modified configurations.

Stealth and persistence

Supply chain attackers aim to remain undetected for as long as possible to maximize the impact of their attack. They may embed backdoors, malware, or other malicious elements within the supply chain, allowing them to maintain persistence within the target's network.

Data exfiltration

Attackers may seek to steal sensitive data, such as intellectual property, customer information, or financial records, from the target organization. This stolen data can be used for various malicious purposes, including espionage or selling on the dark web.

Software supply chain attacks

A common form of supply chain attack involves compromising software updates or packages distributed by trusted vendors. When organizations install these tainted updates, they inadvertently introduce malware or vulnerabilities into their systems.

Hardware supply chain attacks

In some cases, attackers may tamper with hardware components, such as routers or server motherboards, during the manufacturing or distribution process. These compromised components can be used to conduct espionage or intercept sensitive data.

Impersonation and trust exploitation

Attackers may impersonate trusted suppliers or partners to gain access to the target organization's network or information. They exploit the trust relationship between the target and its suppliers to deliver malicious payloads or conduct phishing campaigns.

Mitigation challenges

Detecting and mitigating supply chain attacks can be challenging because they often bypass traditional perimeter security measures. Organizations must implement robust monitoring, threat intelligence, and security assessments throughout their supply chain to mitigate these risks effectively.

Notable examples

High-profile supply chain attacks include the SolarWinds attack, where a compromised software update led to the infiltration of numerous government and private sector organizations, and the NotPetya attack, which disrupted operations for numerous global companies by compromising an accounting software provider's update mechanism.

To defend against supply chain attacks, organizations should conduct risk assessments of their supply chain partners, regularly monitor network traffic and system activity, apply security best practices to their supply chain relationships, and implement multi-layered security measures to detect and respond to potential breaches quickly. Collaboration and information sharing within industries and sectors can also help raise awareness and strengthen collective defenses against these evolving threats.

APT

Imagine a digital adversary that is patient, highly skilled, and relentless – this is the essence of an APT. Think of APTs as the master spies of the cyber world, conducting covert operations with specific objectives. These objectives can range from stealing valuable data and intellectual property to infiltrating the networks of governments, corporations, or organizations for the long haul.

APTs stand out due to their exceptional skill and precision. The cyber operatives behind them are often well-funded and possess an in-depth understanding of computer systems and network vulnerabilities. They craft their attacks meticulously, customizing them for each target, and they are willing to invest months or even years to achieve their goals.

Rather than seeking quick financial gain, APTs focus on the big picture. They're after long-term objectives, which might include espionage, gaining a political or economic advantage, or maintaining persistent surveillance over a target.

To achieve their aims, APTs use a combination of advanced techniques. They might start with spear-phishing emails or compromised websites to gain a foothold within their target's network. Once inside, they move stealthily, evade detection, escalate privileges, and deploy additional malware to keep control.

What makes them particularly challenging is their ability to stay hidden. APTs are like digital ghosts, covering their tracks, erasing logs, and using encryption to cloak their communications. This persistence and stealth are why they're called *advanced* and *persistent*.

Some APTs are believed to be state-sponsored, backed by governments, and motivated by political, military, or economic interests. Others are more profit-oriented, focusing on stealing valuable information for financial gain.

The battle against APTs is a constant and evolving one. Defending against them requires a proactive and multi-layered approach to cybersecurity.

Several APT groups have gained notoriety due to their high-profile attacks, advanced tactics, and persistence. While it's essential to note that attribution in the cybersecurity world can be challenging, experts have identified and tracked various APT groups based on their tactics, techniques, and targets. Here are some of the most famous APT groups:

Name	Attribution	Activities	Objectives
APT29 (Cozy Bear)	Believed to be associated with Russian state-sponsored hacking	Cozy Bear is known for its involvement in the 2016 **Democratic National Committee (DNC)** breach, believed to be part of Russian interference in U.S. elections.	Espionage and information gathering
APT28 (Fancy Bear)	Also linked to Russian state-sponsored hacking	Fancy Bear was implicated in various cyber espionage campaigns, including targeting international organizations, governments, and CI. It was behind the breach of the **World Anti-Doping Agency (WADA)** and the DNC.	Espionage, information theft, and political influence
APT35 (Charming Kitten)	Believed to be associated with Iran	Charming Kitten has targeted political dissidents, journalists, and Middle Eastern governments. It has also been linked to spear-phishing campaigns.	Espionage, surveillance, and political influence

Name	Attribution	Activities	Objectives
APT1 (Comment Crew)	Believed to be linked to the Chinese government's People's Liberation Army Unit 61398	Comment Crew was exposed by Mandiant in 2013 for its extensive cyber espionage operations targeting various industries, including defense, aerospace, and technology companies.	Intellectual property theft and cyber espionage
APT41 (Double Dragon)	Initially associated with Chinese state-sponsored activities but later linked to cybercrime operations	Double Dragon is unique in that it combines state-sponsored and criminal hacking. It has targeted a wide range of industries, including healthcare, gaming, and telecommunications.	Espionage, intellectual property theft, and financial gain
APT33 (Elfin)	Believed to be associated with Iran	Elfin has targeted the energy and aerospace sectors, focusing on CI organizations.	Cyber espionage, targeting CI
APT40 (Temp.Periscope)	Linked to China's state-sponsored hacking activities	Temp.Periscope has targeted naval and defense organizations in Southeast Asia. It is known for its focus on maritime interests.	Espionage, gathering naval intelligence
APT10 (Menupass)	Linked to Chinese state-sponsored hacking	Menupass has targeted **managed service providers (MSPs)** and their customers, including global corporations.	Cyber espionage and intellectual property theft

Table 4.6 – APT groups

These are just a few examples of well-known APT groups, and there are many more active around the world. Each APT group exhibits unique tactics and objectives, making it essential for organizations to stay informed about evolving threats and continually enhance their cybersecurity measures to defend against APT attacks.

Phishing

The world of CI, which includes power grids, water supply systems, transportation networks, and communication infrastructure, is the backbone of modern society. It is also a prime target for cybercriminals seeking to exploit vulnerabilities, often relying on social engineering tactics to gain access. Phishing attacks have proven to be a persistent and effective method for cybercriminals to trick CI personnel into divulging sensitive information.

The anatomy of a phishing attack

Phishing is a type of cyberattack where malicious actors impersonate trustworthy entities, often via email, to deceive individuals into taking harmful actions. In the context of CI, attackers use phishing as a means to infiltrate systems, compromise sensitive data, and potentially disrupt essential services.

Impersonation and trust exploitation

Cybercriminals often craft convincing emails that appear to come from known and trusted sources, such as colleagues, supervisors, or government agencies. These emails may contain urgent requests, seemingly official logos, and professionally written content. CI personnel, who are accustomed to dealing with important matters daily, may be more susceptible to these tactics.

Pretexting and urgency

Attackers frequently create a sense of urgency or emergency within their phishing emails. They may claim that immediate action is required to prevent a disaster, security breach, or data loss. This sense of urgency can cloud judgment and lead recipients to respond hastily without due diligence.

Mimicking authority figures

Phishing emails may impersonate high-ranking individuals within an organization or government agency. This tactic exploits the natural inclination of employees to follow instructions from superiors. CI personnel are often conditioned to obey directives from senior leadership without question.

Deception and lure

Cybercriminals may employ various lures to manipulate recipients. These could include fake invoices, security alerts, job offers, or requests for login credentials. In the context of CI, such lures are crafted to be relevant and convincing to the target audience.

Malicious links and attachments

Phishing emails often contain malicious links or attachments that, when clicked or opened, can execute malware or direct the victim to a fake login page where sensitive credentials are harvested. Attackers may also use weaponized documents that exploit software vulnerabilities.

Why do phishing tactics persist?

Despite increased awareness and cybersecurity measures, phishing attacks, especially in the context of CI, remain prevalent for several reasons.

Human vulnerability

Attackers recognize that humans are often the weakest link in cybersecurity. They exploit psychological traits such as curiosity, trust, and fear to manipulate individuals into taking actions that compromise security.

Spear phishing

Cybercriminals have become adept at spear phishing, a targeted form of phishing where attackers customize their attacks for specific individuals or organizations. By researching their targets, attackers can create highly convincing and tailored phishing attempts that are challenging to detect.

Sophistication and persistence

Phishing attacks have evolved to bypass traditional security defenses. Attackers continually adapt their tactics to stay one step ahead of security solutions.

Financial motivation

The potential for financial gain, whether through ransom payments, data theft, or other cybercrimes, incentivizes attackers to persist in their efforts.

In conclusion, phishing attacks targeting CI personnel continue to pose a significant threat due to the success of social engineering tactics and attackers' adaptability. Vigilance, training, and a multi-layered security approach are essential in defending against these persistent and evolving threats. Protecting CI is not only a matter of cybersecurity; it is crucial to the safety and stability of modern society.

Common unpatched vulnerabilities

In the digital age, the importance of timely patching cannot be overstated. Neglecting to patch systems and applications in a timely manner can have dire consequences, as it opens the door for cyber adversaries to exploit vulnerabilities and infiltrate essential systems. In this section, we will explore the significance of timely patching and shed light on the common unpatched vulnerability exploits that adversaries often leverage to target CI systems.

The significance of timely patching

Timely patching is a fundamental pillar of cybersecurity, especially when it comes to protecting CI. Here are some key reasons why it's essential.

Vulnerability remediation

Patches are released by software and hardware vendors to address known vulnerabilities. Applying patches promptly closes these security holes, reducing the risk of exploitation.

Even though zero-day vulnerabilities (unpatched and unknown vulnerabilities) pose a significant threat, many attackers still rely on known vulnerabilities. Timely patching provides protection against the majority of attacks that exploit known weaknesses.

Data protection

CI systems often handle sensitive data, such as personal information, financial records, or operational data. Neglecting patching can lead to data breaches, compromising the integrity and privacy of this information.

Preventing disruptions

Patching helps maintain system stability and reliability. Neglecting to patch can result in system crashes, downtime, and service disruptions, which can have severe economic and societal consequences.

Legal and regulatory compliance

Many industries and organizations are subject to regulatory requirements that mandate timely patching. Non-compliance can lead to legal penalties and damage to an organization's reputation.

Common unpatched vulnerability exploits

Cyber adversaries often target CI systems by exploiting vulnerabilities that remain unpatched due to neglect. Here are some famous exploits in the last years:

- **EternalBlue (MS17-010):** This well-known exploit targets a vulnerability in the Microsoft Windows **Server Message Block (SMB)** protocol. It was infamously used in the WannaCry ransomware attack, which affected critical systems worldwide.

- **Heartbleed (CVE-2014-0160):** Heartbleed is a severe vulnerability in the OpenSSL cryptographic library. It allows attackers to steal sensitive data, such as SSL/TLS private keys, passwords, or session cookies, compromising the security of encrypted communications.

- **Shellshock (CVE-2014-6271)**: Shellshock is a vulnerability in the Bash shell used in Unix-based operating systems. Attackers can exploit it to execute arbitrary commands remotely, potentially gaining full control of a system.

- **Apache Struts (CVE-2017-5638)**: Exploiting this vulnerability in the Apache Struts framework allows attackers to execute arbitrary code on web servers. It was a significant factor in the Equifax data breach in 2017.

- **BlueKeep (CVE-2019-0708)**: BlueKeep is a remote code execution vulnerability in the Microsoft Windows **Remote Desktop Protocol** (RDP). If exploited, it can lead to unauthorized access and control over a system.

- **Log4J(CVE-2021-44228)**: The Log4j vulnerability is a critical security flaw in the Log4j 2 library which is a popular Java-based logging utility used in many software applications and services. This vulnerability allows for **remote code execution** (RCE) by exploiting the way Log4j 2 processes log messages.

In the world of CI, timely patching is not a mere best practice; it is an imperative. The consequences of neglecting this vital aspect of cybersecurity are far-reaching and can result in grave damage to society, the economy, and individuals. By prioritizing timely patching, organizations and governments can fortify their defenses against the relentless onslaught of cyber adversaries and safeguard the systems that underpin modern life. In a world increasingly reliant on digital infrastructure, the power of timely patching is the shield that keeps the darkness of neglect at bay.

Summary

This chapter offered an in-depth exploration of the most prevalent cyberattacks targeting CI on a global scale. It equipped readers with valuable technical insights into the mechanics of these attacks, shedding light on how they are initiated and executed, and how they ultimately achieve their goals. This chapter covered DDoS attacks, ransomware attacks, supply chain attacks, APT attacks, phishing, and common unpatched vulnerability exploits. It provided readers with a multifaceted skill set, encompassing technical knowledge about different cyberattacks, insight into threat actors' motives, and practical guidance on safeguarding CI against these persistent and evolving threats.

In the upcoming chapter, we'll dive into key real-world incidents of cybersecurity intrusions affecting CI. Building on the foundational knowledge from *Chapters 1–4*, this exploration will deepen your grasp of the cyberattack landscape, equipping you with a more technical perspective.

References

To learn more about the topics that were covered in this chapter, take a look at the following resources:

- *Mansfield-Devine, S. (2014). The evolution of DDoS. Computer Fraud & Security, 2014(10), 15-20.*

- *Zargar, S. T., Joshi, J., & Tipper, D. (2013). A Survey of Defense Mechanisms Against Distributed Denial of Service (DDoS) Flooding Attacks. IEEE Communications Surveys & Tutorials, 15(4), 2046-2069.*

- *Cloudflare. (2023). DDoS threat report for 2023 Q4:* `https://blog.cloudflare.com/ddos-threat-report-2023-q4/`

- *Imperva (2023). The Imperva Global DDoS Threat Landscape Report 2023:* `https://www.imperva.com/resources/reports/the-imperva-global-ddos-threat-landscape-report-2023.pdf?ref=blog.kybervandals.com`

- *MITRE. (n.d.). Technique T1498.002: Network Denial of Service | MITRE ATT&CK®:* `https://attack.mitre.org/techniques/T1498/002/`

- *Cybersecurity & Infrastructure Security Agency. (2014, January 17). UDP-Based Amplification Attacks:* `https://www.cisa.gov/news-events/alerts/2014/01/17/udp-based-amplification-attacks`

- *Cybersecurity & Infrastructure Security Agency. (2013, March 29). DNS Amplification Attacks:* `https://www.cisa.gov/news-events/alerts/2013/03/29/dns-amplification-attacks`

- *Nath, K., Kumar, R.(July 2020).Denial of Service attack using Slowloris. International Research Journal of Engineering and Technology (IRJET):* `https://www.irjet.net/archives/V7/i7/IRJET-V7I779.pdf`

- *Yevsieieva, O., & Helalat, S. M. (2017, October). Analysis of the impact of the slow HTTP DOS and DDOS attacks on the cloud environment. In 2017 4th International Scientific-Practical Conference Problems of Infocommunications. Science and Technology (PIC S&T) (pp. 519-523). IEEE.*

- *Cambiaso, E., Chiola, G., & Aiello, M. (2019). Introducing the SlowDrop Attack. Computer Networks, 150, 234-249.*

- *Ray, M., & Dispensa, S. (2009). Renegotiating TLS.*

- *Bhargavan, K., & Leurent, G. (2016, February). Transcript Collision Attacks: Breaking Authentication in TLS, IKE, and SSH. In Network and Distributed System Security Symposium--NDSS 2016.*

- *Imperial Violet. (2014, October 14). POODLE attacks on SSLv3:* `https://www.imperialviolet.org/2014/10/14/poodle.html`

- *Cybersecurity & Infrastructure Security Agency. (2014, October 17). SSL 3.0 Protocol Vulnerability and POODLE Attack. CISA:* `https://www.cisa.gov/news-events/alerts/2014/10/17/ssl-30-protocol-vulnerability-and-poodle-attack`

- Halfond, W. G., Viegas, J., & Orso, A. (2006, March). A Classification of SQL Injection Attacks and Countermeasures. In Proceedings of the IEEE International Symposium on Secure Software Engineering (Vol. 1, pp. 13-15). Piscataway, NJ: IEEE.

- Sadeghian, A., Zamani, M., & Abdullah, S. M. (2013, September). A Taxonomy of SQL Injection Attacks. In 2013 International Conference on Informatics and Creative Multimedia (pp. 269-273). IEEE.

- Abdullayev, V., & Chauhan, A. S. (2023). SQL Injection Attack: Quick View. Mesopotamian Journal of CyberSecurity, 2023, 30-34.

- Mouli, V. R., & Jevitha, K. P. (2016). Web Services Attacks and Security - A Systematic Literature Review. Procedia Computer Science, 93, 870-877.

- Mainka, C., Jensen, M., Iacono, L. L., & Schwenk, J. (2012). Robust and Effective XML Signatures for Web Services.

- Wendlant, D., Andersen, D. G., & Perrig, A. (2006). Bypassing Network Flooding Attacks using FastPass. Technical report, Carnegie Mellon University.

- Man, K., Qian, Z., Wang, Z., Zheng, X., Huang, Y., & Duan, H. (2020, October). DNS Cache Poisoning Attack Reloaded: Revolutions with Side Channels. In Proceedings of the 2020 ACM SIGSAC Conference on Computer and Communications Security (pp. 1337-1350).

- O'Kane, P., Sezer, S., & Carlin, D. (2018). Evolution of ransomware. IET Networks, 7(5), 321-327.

- Huang, Danny Yuxing, Maxwell Matthaios Aliapoulios, Vector Guo Li, Luca Invernizzi, Elie Bursztein, Kylie McRoberts, Jonathan Levin, Kirill Levchenko, Alex C. Snoeren, and Damon McCoy. "Tracking Ransomware End-to-End." In 2018 IEEE Symposium on Security and Privacy (SP), pp. 618-631. IEEE, 2018.

- Oz, H., Aris, A., Levi, A., & Uluagac, A. S. (2022). A Survey on Ransomware: Evolution, Taxonomy, and Defense Solutions. ACM Computing Surveys (CSUR), 54(11s), 1-37.

- Laszka, A., Farhang, S., & Grossklags, J. (2017). On the Economics of Ransomware. In Decision and Game Theory for Security: 8th International Conference, GameSec 2017, Vienna, Austria, October 23-25, 2017, Proceedings (pp. 397-417). Springer International Publishing.

- Ussath, M., Jaeger, D., Cheng, F., & Meinel, C. (2016, March). Advanced persistent threats: Behind the scenes. In 2016 Annual Conference on Information Science and Systems (CISS) (pp. 181-186). IEEE.

- Rot, A., & Olszewski, B. (2017, September). Advanced Persistent Threats Attacks in Cyberspace. Threats, Vulnerabilities, Methods of Protection. In FedCSIS (Position Papers) (pp. 113-117).

- Khan, S. R., & Gouvia, L. B. (2017). Cybersecurity Attacks: Common Vulnerabilities in the Critical Infrastructure. PASJ International Journal of Computer Science (IIJCS), 5(6), 7-14.

- Arora, A., Forman, C., Nandkumar, A., & Telang, R. (2010). Competition and patching of security vulnerabilities: An empirical analysis. Information Economics and Policy, 22(2), 164-177.

5

Analysis of the Top Cyberattacks on Critical Infrastructure

As we embark on the fifth chapter of our journey through the intricate world of cyberattacks on critical infrastructure, we will go deeper into the heart of the matter – the very real and relentless threats facing our critical infrastructure. Building upon the foundational knowledge acquired in the preceding chapters, this chapter aims to elevate your understanding of cybersecurity to a more technical and sophisticated level.

Having navigated through *Chapters 1* to 4, you've gained a comprehensive understanding of the fundamental principles and concepts that build the world of cybersecurity. Now, it's time to peel back the layers and confront the reality: our critical infrastructure is under constant siege from malicious actors who exploit vulnerabilities with wily precision. In this chapter, we will explore a series of real-world examples, each painting a vivid picture of the havoc these attacks can wreak.

The mission of this chapter is clear: to provide you with a panoramic view of the cybersecurity threat landscape as it pertains to critical infrastructure. We will uncover the methods, motives, and consequences behind these attacks, offering insights that transcend the theoretical and bring you closer to the front line of this ongoing battle.

In this chapter, we will cover the following topics:

- Stuxnet attack on Iran's nuclear program (2010)
- Ukrainian power grid attack (2015)
- DYN attack on internet infrastructure (2016)
- WannaCry (2017)
- NotPetya(2017)
- SolarWinds attack (2020)
- Colonial Pipeline ransomware attack (2021)

By the end of this chapter, you will possess a stronger technical perspective, equipping you with the knowledge and awareness necessary to safeguard vital systems and infrastructure in an increasingly interconnected world. So, fasten your cyber seatbelts as we navigate the tangible, ever-evolving, and often dangerous world of cyber threats against our most critical assets.

Stuxnet attack on Iran's nuclear program (2010)

Stuxnet is a sophisticated worm that emerged in 2010 and is widely believed to be a collaborative effort between multiple nations. Its primary objective was to sabotage Iran's nuclear enrichment program, specifically targeting the centrifuges used to enrich uranium.

Stuxnet specifically targeted **Supervisory Control and Data Acquisition (SCADA)** systems, more specifically the Siemens Step7 software, which is used to program **Industrial Control Systems (ICS)**, including **Programmable Logic Controllers (PLCs)**. The PLCs control the actual machinery in an industrial environment, such as the centrifuges in Iran's nuclear facilities, which Stuxnet targeted.

Stuxnet employed a multi-stage attack process that combined multiple techniques to infiltrate and manipulate ICS. The target was the Siemens SCADA systems used to control the centrifuges at Iran's Natanz uranium enrichment facility.

This is how the attack went down:

Technical Description	
Infection and propagation	Stuxnet initially spread through infected USB drives and network shares. Once on a system, it exploited several zero-day vulnerabilities in Windows and SCADA software to gain access and propagate.
Windows zero-day exploits	**LNK/PIF Shortcut Files Automatic File Execution Vulnerability (CVE-2010-2568)**: The most infamous of the zero-days. Stuxnet used this vulnerability to execute itself via specially crafted Windows shortcut files (.lnk) that it placed on USB drives. When the USB drive was accessed, even just for viewing its contents, the malware would automatically run.
	Print Spooler Service Impersonation Vulnerability (CVE-2010-2729): Stuxnet used this vulnerability to spread to other machines. Once inside a computer, it could use the print spooler service to write itself to other systems on the network.
	Server Service NetPathCanonicalize() Overflow (CVE-2008-4250): While not exactly a zero-day by the time Stuxnet was discovered (since a patch was available by late 2008), Stuxnet still used this vulnerability to propagate itself on networks with unpatched systems.
	Task Scheduler Privilege Escalation Vulnerability (CVE-2010-3338): Stuxnet used this vulnerability to escalate its privileges once on a system.

Technical Description	
Targeted attack on centrifuges	Stuxnet's primary objective was to manipulate the speed of the centrifuges by altering the SCADA systems' programming logic. It targeted the Siemens S7-300 PLCs used in controlling the centrifuges' speed.
Infiltration of SCADA systems via lateral movement	**Profibus network**: Stuxnet scanned the Profibus network, a standard industrial network protocol that many PLCs (including Siemens) use, to identify target devices. **Siemens Step7 project file hijacking**: Stuxnet intercepted and modified the code being uploaded to the PLCs via the Step7 software. By doing this, Stuxnet was able to inject its own code into the PLCs without the operators knowing, causing the centrifuges to act erratically while reporting normal operations. **Siemens default hardcoded credentials**: Stuxnet used hardcoded credentials to access and manipulate the databases associated with the Siemens Step7 software. Specifically, it used these credentials to access the SQL database where project details and configurations were stored. **Siemens S7 communication**: Stuxnet had a deep understanding of the S7 communication protocol used by Siemens PLCs. This knowledge allowed it to intercept and modify communications between the programming software and the PLCs without detection. **Person-in-the-Middle attack on the OPC (OLE for process control) server**: Stuxnet used a person-in-the-middle attack to intercept communications between the PLCs and the Windows-based systems that monitored them. This allowed Stuxnet to report back false information, making it seem like everything was operating normally when, in fact, the machinery was being sabotaged.
Rootkit functionality	Stuxnet utilized rootkit techniques primarily to conceal its activities and maintain persistence on infected systems, making detection and removal particularly challenging.

Technical Description	
Rootkit techniques	**Kernel mode rootkit**: Stuxnet contained a kernel-mode rootkit that allowed it to hide its activities at a very deep level within the operating system. This level of access made it particularly challenging to detect since it could intercept and modify system calls often used by security software for checks.
	Driver signing: To ensure its kernel-mode rootkit was loaded without issues on Windows systems, Stuxnet used stolen digital certificates to sign its drivers, making them appear legitimate. These stolen certificates were from legitimate companies (Realtek and JMicron), thereby bypassing the Windows' requirement that kernel-mode drivers be digitally signed.
	Hiding files and processes: Stuxnet could hide files and processes related to its operations. This made detection by typical means, such as looking at running processes or searching for suspicious files, ineffective.
	File infection: Stuxnet had the ability to infect executable files, ensuring its code would be executed whenever the infected file ran. This also provided a means of persistence, as these infected executables would continue to run the malware even if other components were detected and removed.
	PLC code concealment: One of the more novel rootkit techniques used by Stuxnet was on the PLC side. After modifying the code on Siemens PLCs, it would intercept read requests, returning the original, unmodified code when queried. This meant operators checking the PLC code would see the expected, unaltered code, unaware that the PLC was actually running the malicious Stuxnet-modified version.
	Person-in-the-Middle attack on OPC server: As mentioned previously, Stuxnet used a person-in-the-middle attack to intercept communications between the PLCs and the Windows-based systems that monitored them. This allowed Stuxnet to present false information to monitoring tools, ensuring that the sabotage went unnoticed.
Command and Control	Stuxnet communicated with its **Command and Control** (**C&C** or **C2**) servers to receive updates and new commands. It used various techniques, including encrypted peer-to-peer communication, to avoid detection.

Technical Description	
C&C techniques	**Limited use of C&C servers**: Unlike many other types of malware, Stuxnet did not heavily rely on its C&C servers for its primary mission (sabotaging centrifuges). Instead, the main functionality was built directly into the worm itself, allowing it to operate even if its C&C infrastructure was taken down or disconnected.
	Domain Generation Algorithm (DGA): Stuxnet utilized a DGA to produce a list of domains for C&C. DGAs generate a large number of possible domain names and are typically used to provide resilience against takedowns. If one domain is taken down, the malware can attempt to contact another.
	Fallback domains: Along with the DGA, Stuxnet had hardcoded fallback domains. If it couldn't connect to the domains generated by the DGA, it would attempt to connect to these fallback domains.
	Peer-to-peer updates: Stuxnet had the capability to update other Stuxnet-infected machines within the same local network. If one infected machine managed to contact a C&C server and receive an update, it could then distribute that update to other infected machines without them having to connect directly to the C&C server.
	Encrypted communications: Communications between Stuxnet-infected machines and the C&C servers were encrypted, making network-based detection and analysis more difficult.
	Stolen certificates: As previously mentioned in the context of the rootkit techniques, Stuxnet used stolen digital certificates. This not only helped in loading its drivers but also in securing and legitimizing its communications.
	Data exfiltration: While Stuxnet's primary goal was sabotage, it was also designed to exfiltrate information about the infected environment. This information could be sent back to the C&C servers, providing the attackers with valuable intelligence about the success of the infection and the nature of the environment it had penetrated.
Manipulation of frequency converters	Stuxnet manipulated the frequency converter devices connected to the centrifuges. It sent rogue commands that caused the frequency converters to oscillate, leading to excessive wear and tear on the centrifuges, ultimately causing them to fail.
Consequences	Stuxnet succeeded in causing physical damage to a significant number of Iran's centrifuges. Its complex design, use of multiple zero-day vulnerabilities, and highly targeted approach made it one of the most remarkable and impactful cyberattacks in history.

Table 5.1 – Stuxnet attack on Iran's nuclear program

The Stuxnet attack is arguably one of the most significant cyber operations in history, not only due to its technical sophistication but also because of its impact on real-world infrastructure and its geopolitical implications. Here are some of the key consequences of the Stuxnet attack:

- Stuxnet targeted and successfully damaged centrifuges in Iran's Natanz nuclear facility by causing the centrifuges to spin out of control while simultaneously displaying normal operating conditions to monitoring systems. Stuxnet led to the physical destruction of a significant portion of Iran's centrifuges.

- Prior to Stuxnet, many cyberattacks focused on data theft or espionage. Stuxnet demonstrated that cyber weapons could cause physical damage to critical infrastructure. This realization shifted global perspectives on the potential impacts of cyber warfare.

- The fact that Stuxnet was discovered and dissected showed that even the most sophisticated cyber operations can be exposed. This offered lessons to nations and organizations about the risks of deploying cyber weapons, as once discovered, their tactics can be analyzed and potentially used by others.

- Post-Stuxnet, several pieces of malware believed to be derived from Stuxnet or developed by the same entities emerged. Examples include Duqu (focused on espionage) and Flame (an advanced spying tool). Additionally, Stuxnet's public dissection meant that cybercriminals could study its techniques, leading to concerns about its tactics being used in other malicious campaigns.

In summary, Stuxnet's consequences extended far beyond the immediate damage it caused to Iran's nuclear program. It reshaped perceptions of cyber warfare, influenced global cybersecurity strategies, and prompted new discussions about cyber ethics and norms.

Ukrainian power grid attack (2015)

The Ukrainian power grid attack, which notably occurred in December 2015, was a significant cyber-physical incident that disrupted Ukraine's power distribution. It was one of the first public examples where a cyber attack led to real-world consequences by causing a widespread power outage.

Here's a technical description of how the attack unfolded:

Technical Description	
Initial compromise	The attackers began with a spear-phishing campaign targeting employees of three Ukrainian regional power distribution companies. The emails contained Microsoft Office documents embedded with the BlackEnergy 3 malware. Once the documents were opened, the malware was delivered to the host computer. Once BlackEnergy was in the target's system, it facilitated the delivery of KillDisk and other modules. KillDisk was used to wipe or corrupt various types of files on infected systems. It also rendered certain machines unbootable by overwriting their **master boot records (MBRs)**.
BlackEnergy	BlackEnergy is a family of malware that has been in existence since at least 2007. It started as a simple toolkit for creating botnets, used mainly for **Distributed Denial of Service (DDoS)** attacks, but has evolved over the years to include a variety of functionalities. BlackEnergy 3 is a more recent and sophisticated version of the malware. BlackEnergy 3 contains a component called **KillDisk**. This module is capable of wiping files and making systems unbootable by overwriting the MBR and partition table. BlackEnergy 3 showed a specific interest in ICS and SCADA environments. It had modules designed to harvest data about industrial control systems, suggesting that its operators had a particular interest in critical infrastructure.
Establishing a foothold	With the initial compromise achieved, the attackers used the malware to harvest credentials, escalate privileges, and move laterally within the victim networks. They utilized common tools such as PsExec and Mimikatz to execute processes remotely and extract credentials from memory, respectively.

Technical Description	
PsExec	PsExec allows administrators to remotely execute commands on a system.
	One of the key features of PsExec is that it does not require any software or agent to be installed on the remote system. It operates over the **Server Message Block (SMB)** protocol, which is natively supported by Windows systems. To use PsExec, administrative credentials are required. If these credentials are compromised, an attacker can use PsExec to execute commands on any machine where those credentials are valid.
Mimikatz	Mimikatz is a well-known post-exploitation tool. It is especially recognized for its ability to extract plaintext passwords, hashes, PIN codes, and Kerberos tickets from memory (specifically from lsass.exe, the Local Security Authority Subsystem Service). This makes it a potent tool in the hands of both system administrators looking to validate security measures and cyber adversaries aiming to exploit compromised systems further.
	Mimikatz can retrieve plaintext passwords stored in memory.
	It also enables attackers to use a password hash for authentication, bypassing the need for the actual password.
	It creates a forged Kerberos **Ticket Granting Ticket (TGT)**, allowing unauthorized access to any account on the domain.
	It creates a forged Kerberos service ticket for specific services on a target machine.
	It uses AES keys to obtain TGTs from the **Key Distribution Center (KDC)** without requiring a password and can extract credentials from the memory of lsass.exe.
Reconnaissance	The attackers spent time observing the operations of the power grid and learning about the ICS environment and SCADA systems.
	The cybercriminals used VPN connections, remote desktop sessions, and admin tools to move around the compromised networks.

Technical Description	
Disrupting the power supply	The attackers manually opened breakers in substations to interrupt the power supply. They achieved this by using the SCADA systems' own controls. It's worth noting that this kind of manual intervention suggests a certain level of familiarity with the victim's operational environment.
Disabling IT infrastructure	The attackers used KillDisk to disable the SCADA systems, thereby preventing operators from gaining control. Workstations and servers were also targeted, rendering many of them inoperative.
Telephone denial-of-service	The attackers also executed a telephonic denial-of-service attack on the utility's customer call centers to prevent customers from reporting or inquiring about the outages.
Recovery and impact	Power was restored manually by dispatching engineers to the affected substations. It's estimated that around 230,000 people were without power for several hours due to the attack.

Table 5.2 – Ukrainian power grid attack

After the 2015 attack, there were more cyber incidents targeting Ukrainian infrastructure, emphasizing the evolving nature of cyber threats and the importance of constant vigilance in protecting critical systems.

Dyn attack on internet infrastructure (2016)

The Dyn attack of 2016 stands out as a landmark event in the annals of cyber warfare, marking one of the most significant DDoS attacks ever recorded. This meticulously orchestrated offensive targeted Dyn, a major US-based **Domain Name System (DNS)** provider, crucial for translating website names into IP addresses. The attack's impact was far-reaching and unprecedented, severely disrupting access to a multitude of popular websites and online services. High-profile sites including Twitter, Netflix, and Reddit were among those affected, highlighting the attack's extensive reach. The incident not only demonstrated the vulnerabilities inherent in digital infrastructure but also underscored the potential for widespread disruption when key internet nodes are compromised.

This is how it worked:

Technical Description	
Initial compromise	The Dyn attack largely utilized Mirai, malware that transforms Linux-operating network devices into bots that can be remotely controlled. These bots become components of a larger botnet, which can be deployed in extensive network attacks. The Mirai malware scans the internet for vulnerable devices, particularly **Internet of Things (IoT)** devices such as security cameras, DVRs, and routers.
Exploitation	Mirai-infected devices begin by scanning the internet for IP addresses of vulnerable IoT devices. The primary targets are devices that listen on telnet ports (23 and 2323), although other ports can be targeted too. Once a target is identified, Mirai attempts to establish a connection using a predefined list of common default usernames and passwords. Many IoT devices are shipped with default login credentials, and users often don't change them, leaving devices vulnerable.
Infection	If Mirai successfully logs in using one of the default credential pairs, it establishes a connection and sends information about the vulnerable device back to a reporting server. The malware then instructs the compromised device to download and execute the Mirai binary. This binary is tailored to the specific architecture of the device.
Botnet recruitment	Once the binary is executed, the device becomes part of the Mirai botnet. It starts to communicate with the C2 server and can receive commands.
C2 server interaction	Each infected device (bot) periodically contacts a C2 server. This server provides instructions to the bots and can command them to launch attacks on target networks.
Launch of the DDoS attack	The attacker instructs the Mirai botnet (which, by the time of the Dyn attack, had amassed a large number of infected devices) to target Dyn's infrastructure. The botnet starts sending a massive volume of requests to Dyn's DNS servers.
First-wave attack	Beginning at approximately 7:00 AM (ET), this was a large-scale attack mainly targeting Dyn's East Coast data centers. As a result, many users on the East Coast of the U.S. experienced difficulty accessing several major websites.
Second wave	A more massive wave started around midday. It was broader, aiming to target both the East Coast and other infrastructures globally. Dyn's countermeasures and increased traffic capacity managed to mitigate this wave within an hour.

Third wave	In the late afternoon, Dyn observed a third wave of attacks, but they had already implemented defensive measures to fend it off, making this wave less disruptive.
Impact on Dyn's services	Dyn's DNS infrastructure was overwhelmed by the sheer volume of malicious requests. Due to the attack on Dyn's systems, legitimate requests from users trying to access websites that used Dyn's DNS service (such as Twitter, Reddit, Netflix, and many others) were delayed or lost, resulting in inaccessibility for many users.

Table 5.3 – Dyn attack on internet infrastructure

It's important to note that the Dyn attack underscored the vulnerabilities of the modern internet architecture, particularly when a significant portion of the web relies on a single service (in this case, Dyn's DNS services). It also highlighted the risks associated with insecure IoT devices. After the attack, there were increased calls for better security standards for IoT devices to prevent similar attacks in the future.

WannaCry (2017)

WannaCry was a ransomware attack that struck globally in May 2017, affecting more than 200,000 computers across 150 countries in just a few days. The attack exploited vulnerabilities in Windows operating systems to spread and encrypt files on infected machines.

Here's a description of the WannaCry attack:

Technical Description	
Initial compromise	The initial misconception about WannaCry's spread was that it was primarily through phishing emails. However, as further analysis took place, it became clear that the primary mechanism of the rapid global propagation was the exploitation of an SMB vulnerability (CVE-2017-0144) via the EternalBlue exploit of systems that had an open (Server Message Block) port (specifically, port 445) to the public internet.
CVE-2017-0144	CVE-2017-0144 is a specific identifier for a vulnerability within the Windows SMB version 1.0. The SMB is a network file-sharing protocol that allows for computer-to-computer communication. If successfully exploited, this vulnerability allows attackers to perform remote code execution on the affected system, meaning they can run arbitrary code and potentially gain full control over the system. This could be achieved without any user interaction if the attacker sends a specially crafted packet to a targeted SMBv1 server.

Technical Description	
The Shadow Brokers	The Shadow Brokers are a mysterious hacking group that came into the limelight in mid-2016. Their exact origins, motivations, and affiliations remain a subject of speculation. The Shadow Brokers announced that they had stolen a cache of cyber weapons and tools from the Equation Group, which is widely believed to be a cyber espionage group linked to the U.S. **National Security Agency (NSA)**. They provided samples and later attempted to auction the entire set._blank_ Among the tools leaked by the Shadow Brokers, EternalBlue stood out. It is an exploit that targets the Microsoft Windows SMB version 1.0 protocol, allowing for unauthorized remote code execution and use in the WannaCry attack.
EternalBlue to gain access	EternalBlue exploits the CVE-2017-0144 vulnerability in Microsoft's implementation of the SMBv1 protocol. Specifically, the exploit targets the way SMBv1 handles certain types of packets and sends specially crafted packets to the SMBv1 server. The packets are constructed in a way that they overflow a buffer in the `srvnet.sys` process of the Windows OS. The primary technical aspect of the exploit is a buffer overflow condition. This allows the attacker to overwrite memory in the targeted system's kernel space. After successfully triggering the buffer overflow, EternalBlue proceeds to execute its shellcode. This shellcode is responsible for loading and running the main payload on the targeted machine. Often used in conjunction with EternalBlue is DoublePulsar, a kernel-level backdoor.
DoublePulsar for persistent access	DoublePulsar is a backdoor implant tool that was also leaked by the Shadow Brokers in their disclosure of alleged NSA tools._blank_ While EternalBlue acted as the primary exploit to gain unauthorized access to systems by targeting the SMB vulnerability, DoublePulsar was commonly used as the subsequent payload to maintain that access and provide further control over the compromised system._blank_ DoublePulsar provides a mechanism to inject malicious DLLs into user-space processes, thus allowing the execution of arbitrary code. It operates at the kernel level, making its activities stealthy and difficult to detect.

Technical Description	
LAN scanning capabilities	Once a machine was infected, WannaCry identified the local IP address and subnet of the compromised machine. This allowed it to determine which IP addresses belonged to the same local network. WannaCry then began scanning the local subnet for other machines, specifically targeting TCP port 445, which is used by the SMB protocol.
	The primary objective was to identify systems that had SMBv1 enabled and were vulnerable to the EternalBlue exploit. When a potential target was identified within the LAN, WannaCry attempted to exploit the SMBv1 vulnerability using the EternalBlue exploit. If the exploitation attempt was successful, the ransomware payload was delivered to the newly compromised machine, and the encryption routine began on that system. The newly infected machine would then repeat the process.
Internet-wide scanning capabilities	In addition to the LAN scanning, WannaCry also initiated a broader scan, randomly targeting IP addresses on the internet. This allowed it to jump between unconnected networks and spread globally.
	This external scanning was more shotgun-style, with a wider and less targeted approach than the internal LAN scanning. Nevertheless, it was still effective, especially given the vast number of exposed and vulnerable systems on the internet.
File encryption	The WannaCry ransomware, once executed on a system, primarily aimed to encrypt the user's files, making them inaccessible until a ransom was paid. WannaCry had a list of specific file extensions it targeted for encryption, which included common data file types such as documents (`.doc`, `.docx`), databases (`.sql`), images (`.jpg`, `.png`), videos (`.mp4`), and many others. This allowed it to focus on files most likely to be of value to the victim.
	For each file, WannaCry generated a random AES key. This key was used to encrypt the actual file content. AES is a symmetric encryption method, meaning the same key can both encrypt and decrypt the data. This allows for faster encryption of large files.
	The randomly generated AES key for each file, after being used, was then encrypted with WannaCry's RSA public key. RSA is an asymmetric encryption algorithm, meaning it has a public key (for encryption) and a private key (for decryption). Only someone with the private RSA key could decrypt the AES keys tied to each file.
	The encrypted AES key was appended to the encrypted file as the attackers would need to decrypt this AES key first (using their private RSA key) if the ransom was paid and decryption was to be provided.
	Encrypted files had their extensions changed to `.wcry`.

Technical Description	
Ransom note display	After the encryption process, WannaCry dropped a ransom note named `@WanaDecryptor@.exe` on the desktop and in folders containing encrypted files. This executable, when run, would display the ransom message, instructing victims on how to pay to recover their files. The ransomware also changed the desktop wallpaper to a message informing the user that their files were encrypted and directing them to the ransom note for details.
Shadow copies	To reduce the chance of victims recovering files without paying the ransom, WannaCry attempted to delete Windows shadow copies (backup or snapshot copies of files) using the `vssadmin` command. This made it more challenging for users to restore their files from system backups.
Kill switch domain	Upon execution, before initiating its encryption routine, WannaCry would attempt to make an HTTP request to a specific, hardcoded domain. This domain was a long and seemingly random string of characters followed by `.com` (`iuqerfsodp9ifjaposdfjhgosurijfaewrwergwea.com`). The domain was not registered at the time of the initial WannaCry outbreak. If the domain did not resolve (that is, if there was no server to respond to the request because the domain wasn't registered), WannaCry would proceed with its malicious activities, including encrypting the victim's files. However, if the domain did resolve and responded to the request, the ransomware would terminate itself without causing any harm. Essentially, the successful connection to the domain acted as an *off switch* for the ransomware.
Purposes of the kill switch	One theory is that the kill switch was an anti-analysis or anti-sandboxing technique. Some malware analysis environments (sandboxes) are designed to simulate internet connectivity by resolving all domain requests, regardless of whether the domains are real. By checking for the resolution of a non-existent domain, WannaCry might have been trying to detect if it was running in a sandbox environment. Another theory is that it was a safety feature inserted by the ransomware's developers. If they ever wanted to halt the spread of the ransomware for any reason, they could simply register the domain themselves.

Technical Description	
Accidental neutralization of WannaCry	A security researcher named Marcus Hutchins (also known by his alias *MalwareTech*) discovered this unregistered domain while analyzing the ransomware's code. Curious about its purpose, he registered the domain, effectively creating a *sinkhole*.
	By registering the domain and allowing it to resolve, Hutchins inadvertently activated the *kill switch*, halting the spread of the initial WannaCry variant. Machines infected after the domain registration would contact the domain, receive a response, and terminate the ransomware process before any damage could occur.

Table 5.4 – WannaCry attack

WannaCry's impact on critical infrastructure was particularly concerning because these systems are vital for the functioning of societies and economies.

Arguably the most high-profile and concerning impact was on the **National Health Service (NHS)**. Over a third of NHS trusts in England were disrupted by the ransomware. Some hospitals had to divert emergency patients to other unaffected facilities. Non-critical appointments and procedures were postponed or canceled. The ransomware hindered access to patient records, affecting the diagnosis and treatment procedures. The attack cost the NHS an estimated £92 million, factoring in the IT response, lost output, and restoring systems and data.

Deutsche Bahn (German Railways) ticketing systems and display panels at train stations were affected by WannaCry, leading to confusion and disruption in train services.

The car manufacturing alliance Renault-Nissan faced disruptions due to WannaCry. Some of its manufacturing plants, notably in France, Romania, and Japan, had to halt operations temporarily to deal with the ransomware's impact.

One of Spain's largest telecommunications providers, Telefónica, was hit hard by WannaCry. While not *critical infrastructure* in the strictest sense, any disruption to a major telecom provider can have downstream effects on other critical services, including emergency communications.

NotPetya (2017)

NotPetya is one of the most notorious cyberattacks in history, believed to have been initiated by a state-sponsored actor. The malware rapidly spread across the globe, affecting thousands of computers in numerous organizations, and causing significant disruptions.

The malware was named **NotPetya** because, at first glance, it appeared to be a variant of the *Petya* ransomware, which had been previously identified and studied by cybersecurity experts. Petya was known for encrypting the MBR of infected systems, preventing them from booting up, and then demanding a ransom.

However, as researchers studied this new variant more, they found significant differences in its operation and intent. While Petya was a genuine ransomware that provided victims with the possibility (though not a guarantee) of decrypting their files after paying a ransom, NotPetya was designed more as a *wiper*, with the primary objective of causing destruction and disruption. Its encryption was done in such a way that data recovery was nearly impossible, even if a ransom was paid.

This is how it worked:

Technical Description	
Initial compromise	NotPetya was initially spread through a compromised update of the MEDoc software, a popular Ukrainian tax accounting package. Attackers gained unauthorized access to MEDoc's update server and pushed a malicious update containing the NotPetya payload to its users.
Payload execution	Once the malicious MEDoc update was installed, the NotPetya payload was executed, first checking for the presence of a *kill switch* file. If the malware detected its own presence (by checking for a specific file), it would not proceed further on that system. Otherwise, it proceeded to the encryption and spreading phase.
Credential harvesting using Mimikatz	NotPetya incorporated a version of the Mimikatz tool to extract passwords from memory. This allowed the malware to acquire user credentials from the infected system.
MBR overwriting	NotPetya overwrote the MBR of the infected computer. This made the system unbootable and displayed a ransom note to the user.
File system encryption	While it appeared to be ransomware, NotPetya was more of a wiper. It encrypted parts of the file system using the *Salsa20* algorithm, making file recovery difficult if not impossible.

Technical Description	
Lateral movement	NotPetya used multiple mechanisms to spread within a network: • EternalBlue and EternalRomance, SMB exploits that were leaked from the NSA: NotPetya weaponized them to spread to other systems in the same network. • WMIC and PsExec, with harvested credentials: NotPetya used **Windows Management Instrumentation Command-line (WMIC)** and PsExec tools to execute the malware on remote systems. Using obtained credentials, it tried to copy itself to the `admin$` share on other machines.
Spread to the world	Although the initial focus was Ukrainian businesses, the malware quickly spread to other organizations connected to these businesses, including their international partners, clients, and other entities. Given the interconnectedness of modern businesses and supply chains, this meant that NotPetya could spread beyond its initial geographic and sectoral targets, impacting organizations worldwide.
Protecting itself	NotPetya attempted to make analysis and mitigation more difficult by disabling several Windows services, including Windows Update, and by shutting down important processes such as antivirus solutions.
Communication with C2 servers	NotPetya didn't maintain an active connection with its C2 servers in the same way traditional malware might. The primary objective appeared to be destruction rather than data exfiltration or remote control.
Decryption improbability	Victims were instructed to pay a ransom in Bitcoin in exchange for a decryption key. However, NotPetya's poor implementation of its payment and decryption process, coupled with its primary objective of destruction, made decryption and recovery highly unlikely even if the ransom was paid.

Table 5.5 – NotPetya attack

NotPetya had significant consequences for critical infrastructure around the world. While the malware was initially targeted at Ukrainian businesses, its rapid spread meant that numerous global organizations were impacted.

Maersk, one of the world's largest shipping companies, was heavily affected. Their operations at multiple ports around the world were disrupted. Maersk later reported that the incident cost them between $250 million and $300 million.

Merck, a major American pharmaceutical company, was hit hard. The disruption affected its manufacturing, research, and sales operations. The financial impact was estimated at around $870 million.

Rosneft, a leading Russian oil producer, reported that they were affected, though the extent of the damage wasn't fully detailed.

Mondelez International, the parent company of brands such as Cadbury, was affected, with disruptions to their shipping and invoices. They reported a 3% drop in sales due to the attack.

FedEx's European subsidiary, TNT Express, was severely impacted. Many of their operations, including shipping and communications, were disrupted. FedEx estimated the cost at roughly $300 million.

Several healthcare institutions were impacted, causing disruptions in services and patient care. For instance, in the U.S., some hospitals had to reschedule surgeries and other critical services.

In Ukraine, where the attack was primarily focused, there was disruption in the power grid. Kyivenergo, a Ukrainian energy company, had to process payments manually due to the attack.

Several banks in Ukraine faced disruptions in their operations, affecting customer transactions and services. Also, public infrastructure such as airports and metro services reported problems. For instance, the capital's Boryspil Airport faced some operational disruptions.

SolarWinds attack (2020)

SolarWinds is an American company that develops software for businesses to help manage their networks, systems, and information technology infrastructure. Founded in 1999 and headquartered in Austin, Texas, the company offers a variety of products designed to help IT professionals monitor and manage the performance of their networks, servers, applications, databases, and other IT infrastructure components.

The SolarWinds cyberattack was one of the most sophisticated and wide-reaching supply chain attacks in history, primarily targeting U.S. government agencies and numerous companies around the world. This attack was attributed to Russian state-sponsored actors, according to U.S. intelligence agencies.

This is how it developed:

Technical Description	
Initial compromise	Attackers gained access to the environment of SolarWinds.
	The exact initial intrusion vector remains uncertain, but sophisticated spear-phishing or exploiting vulnerabilities are common methods.

Technical Description

Inserting malicious code into the Orion Platform	Once inside, the attackers inserted malicious code into the source code of the SolarWinds Orion Platform, a widely used network monitoring and management tool. This malicious code was designed to be stealthy, with functions and behaviors mimicking legitimate SolarWinds code to avoid detection.
Build process	The malicious code was then included in the legitimate build process of the Orion software. As a result, official software updates delivered by SolarWinds to its customers contained a malicious backdoor. This type of attack is termed a *supply chain attack*. Because the malicious code was part of the official build process, the resulting software binaries were signed with a valid SolarWinds digital certificate, making the update appear legitimate to end users and most security tools.
Activation	Orion Platform versions 2019.4 through 2020.2.1, released between March 2020 and June 2020, were compromised. Organizations that used these versions were potentially affected by the SUNBURST (or Solorigate) backdoor. The primary malicious **dynamic link library** (**DLL**) file associated with the attack was named `SolarWinds.Orion.Core. BusinessLayer.dll`. This DLL was a component of the Orion software. When this tainted DLL was executed, it resulted in the SUNBURST backdoor being activated, which subsequently allowed the attackers to conduct their operations.
Command and Control (C&C)	Once activated, the SUNBURST backdoor would make an initial C2 communication to a subdomain of `avsvmcloud[.]com`. Upon successful communication with the C2 server, the malware began its stealthy reconnaissance phase, gathering details about the infected environment. This included system configurations, user accounts, network architecture, and more. It also checked for security products and services to potentially avoid detection. The attackers used methods to harvest credentials, leveraging tools such as Mimikatz, a password dumping tool, or custom scripts to extract them from memory, disk, or configuration files. In some cases, the attackers also created new accounts or leveraged compromised accounts, giving them more flexibility to move around.

Technical Description	
Lateral movement and further compromise	The attackers used a combination of native Windows tools and custom utilities. For example, they leveraged PowerShell scripts and utilities such as PsExec to execute commands remotely on other machines. **Windows Management Instrumentation (WMI)** was also used to execute commands or scripts on remote computers. The goal here was to gain higher-level privileges if the initially compromised account didn't have them. The attackers looked for vulnerabilities or misconfigurations in the system that could be exploited to elevate their privileges. By extracting tokens from logged-in or recently logged-out users, the attackers could impersonate those users or roles, allowing them to access resources and carry out tasks as if they were those users.
Establishing persistence	Apart from the SUNBURST backdoor, the attackers deployed other tools to ensure they maintained access. TEARDROP and RAINDROP are examples of custom malware loaders used in some instances. They also used *Golden Tickets* (Kerberos ticket-granting tickets) and *Silver Tickets* (service-specific Kerberos tickets) for persistence and further movement.
Data exfiltration	Targeted data was exfiltrated from the compromised networks. The attackers went to great lengths to camouflage their data theft, sometimes even using trusted third-party **content delivery networks (CDNs)** and storage services.
Discovery and mitigation	In December 2020, the breach was publicly disclosed after FireEye, a leading cybersecurity firm (and a victim itself), detected the anomaly. Affected organizations, with the assistance of cybersecurity experts and vendors, began the complex task of investigating, eradicating the malware, and securing their networks.
Consequences	Several U.S. government agencies were compromised, including parts of the Department of Defense, the Department of Homeland Security, the Department of the Treasury, the Department of State, the Department of Energy, and the National Nuclear Security Administration. This raised concerns about potential espionage, theft of sensitive data, and potential disruptions to critical operations.

Table 5.6 – SolarWinds attack

The SolarWinds attack had far-reaching implications, necessitating a reevaluation of national cybersecurity policies and demanding stricter oversight of software supply chains.

In the wake of the breach, there were strong calls for enhanced cybersecurity measures, with a particular emphasis on adopting zero-trust architectures, prioritizing continuous monitoring, and ensuring transparency. Notably, even industry giants like FireEye were not immune, with their internal investigations revealing the extent of the attack.

This incident highlighted the pressing need for a shift in security perspectives, from signature-based threat detection to behavior-based security, and drew renewed attention to potential vulnerabilities such as insider threats.

Colonial Pipeline ransomware attack (2021)

The Colonial Pipeline ransomware attack occurred in May 2021, impacting one of the largest fuel pipeline systems in the United States. A ransomware group known as *DarkSide* was responsible for the attack.

The Colonial Pipeline is a significant piece of energy infrastructure in the United States. Stretching over 5,500 miles, the pipeline system spans from Houston, Texas, on the Gulf Coast, to the New York Harbor area. The pipeline can transport more than 3 million barrels of gasoline, diesel fuel, jet fuel, and other refined products per day. It services seven airports and numerous military installations.

The Colonial Pipeline plays a crucial role in supplying fuel to the U.S. East Coast, servicing an estimated 50 million consumers daily. It supplies roughly 45% of the fuel consumed on the East Coast, making it a critical piece of infrastructure for the region's energy security and economic stability.

The attack on the Colonial Pipeline involved a ransomware infection targeting the pipeline's IT systems and business operations. The specific technical details have not been publicly disclosed in full.

This is what we know:

Technical Description	
Initial compromise	The attackers exploited an old VPN account that was no longer in use but was still active within the Colonial Pipeline's network. This VPN was designed to allow employees to access the company's IT systems remotely and securely. The password for the VPN account was found among a batch of leaked passwords on the dark web. It's unclear how the password initially got leaked or whether it was part of a larger data breach.
Lateral movement	The attackers then moved laterally across the network, searching for valuable data and the infrastructure responsible for the Colonial Pipeline's operations.

Technical Description	
Data exfiltration	Data theft or exfiltration often precedes ransomware attacks. In the case of the Colonial Pipeline attack, the culprits managed to steal approximately 100 gigabytes of data. This operation was conducted swiftly, within a two-hour window, indicating that the attackers had a clear understanding of the network and where the valuable data resided.
Deploying ransomware	After identifying their targets within the network, the attackers deployed the DarkSide ransomware. DarkSide is **Ransomware-as-a-Service** (**RaaS**), where the malware's developers provide the ransomware to affiliates who carry out attacks and share a portion of the ransom payment with the developers. The ransomware encrypted files on affected systems, rendering them inaccessible. A ransom note would be displayed demanding payment for decryption.
DarkSide collecting information	DarkSide collects the computer's hostname and the current user's username. This information can give insights into whether the infected machine is a personal computer or belongs to an organization, as server or workstation naming conventions might reveal the machine's purpose. The ransomware gathers details about the operating system, such as the OS version, architecture (32-bit or 64-bit), and build number. This can be achieved using system API calls or by querying system information. DarkSide checks the system's language and regional settings. If it detects certain languages, especially those from former Soviet bloc nations, it will terminate its operations. The ransomware enumerates active processes and services on the system to identify and terminate security or backup-related processes that could hinder its operations and to gain insights into the system's role and the type of software it's running, which could indicate the value of the machine to the organization. The ransomware also collects network details to prepare for the attack.
DarkSide sending data to Command and Control servers	Once the relevant system information is gathered, DarkSide typically sends it back to its C2 servers. This communication can be achieved using HTTP/HTTPS requests, custom protocols, or other means. The data can be encrypted or obfuscated to avoid detection during transmission.

Technical Description	
DarkSide selecting files for encryption	DarkSide has a list of specific file extensions that it targets for encryption. These usually include document files (for example, `.docx`, `.pdf`), database files (for example, `.mdb`, `.db`), image files (for example, `.jpg`, `.png`), and other types of commonly used files that contain valuable information.
	The ransomware also has a list of file extensions and directories it intentionally avoids, such as system-critical file extensions (`.exe`, `.dll`, `.sys`, and others) that are necessary for the OS to function correctly. Encrypting these files could make the system unbootable, which would be counterproductive for the ransomware's goal of displaying the ransom note and getting paid.
	Directories such as `Windows`, `Program Files`, and other system-related folders are typically excluded to prevent system corruption.
	While not directly related to file selection for encryption, it's worth noting that DarkSide, like many other ransomware strains, tries to delete Volume Shadow Copies on Windows systems. These are backup snapshots that can potentially be used to restore encrypted files. By deleting them using tools such as `vssadmin.exe`, the ransomware ensures that victims have fewer avenues to recover their data without paying the ransom.
	As part of its RaaS model, DarkSide might allow its affiliates some level of customization in targeting files. Depending on the intended victim or the specific campaign, the list of targeted or excluded files might be adjusted.

Table 5.7 – Colonial Pipeline ransomware attack

The attack had a significant impact on Colonial's business operations. As a precautionary measure, Colonial shut down its pipeline operations to contain the spread of the ransomware and to ensure the integrity of its **operational technology (OT)** systems.

This shutdown led to widespread fuel shortages and panic buying in many parts of the U.S. Eastern Coast.

Colonial engaged external cybersecurity firms to investigate and respond to the incident. They also contacted law enforcement agencies, including the FBI.

The company reportedly paid a ransom of nearly $5 million to the attackers to obtain the decryption key and restore their systems.

With the decryption key and the assistance of cybersecurity experts, Colonial began the process of restoring its IT systems and, subsequently, its pipeline operations.

This recovery process involves ensuring that no remnants of the malware remain, patching vulnerabilities, and strengthening security measures to prevent future attacks.

Summary

As the digital era unfolds, the boundaries between our electronic existence and tangible reality have become increasingly indistinct. We witnessed this decade's evolving shadows of cyber warfare: the eerie silence after Stuxnet's attack, the chilling darkness following Ukraine's power grid sabotage, and the chaos borne from the ransom waves of WannaCry and NotPetya. The very tools we place our trust in were exposed in the SolarWinds incident, while episodes involving the Colonial Pipeline remind us how precariously our daily lives hang in the balance.

In this chapter, we ventured deep into the maze of cyber conflict, spotlighting some of the most consequential attacks that have shaken our world's foundation. As we hurtle forward in this digital age, the repeated sieges on our essential infrastructures bring to light a pressing concern: the indispensable need to bolster our defenses. Let the tales of these breaches serve as both a warning and a guide, helping us navigate and protect our interconnected future.

In the next chapter, we'll learn about critical aspects of protecting critical infrastructure, focusing on the implementation of security policies and frameworks, enhancing network security, and the importance of continuous monitoring in the face of growing cyber threats.

References

To learn more about the topics that were covered in this chapter, take a look at the following resources:

- *Farwell, J. P., & Rohozinski, R. (2011). Stuxnet and the future of cyber war. Survival, 53(1), 23-40.*

- *Matrosov, A., Rodionov, E., Harley, D., & Malcho, J. (2010). Stuxnet under the microscope. ESET LLC (September 2010), 6.*

- *Kushner, D. (2013). The real story of stuxnet. ieee Spectrum, 50(3), 48-53.*

- *Karnouskos, S. (2011, November). Stuxnet worm impact on industrial cyber-physical system security. In IECON 2011-37th Annual Conference of the IEEE Industrial Electronics Society (pp. 4490-4494). IEEE.*

- *Lindsay, J. R. (2013). Stuxnet and the limits of cyber warfare. Security Studies, 22(3), 365-404.*

- *Langner, R. (2011). Stuxnet: Dissecting a cyberwarfare weapon. IEEE Security & Privacy, 9(3), 49-51.*

- *Dudley, R., & Golden, D. (2021). The colonial pipeline ransomware hackers had a secret weapon: self-promoting cybersecurity firms. ProPublica (24 May 2021).*

- *Beerman, J., Berent, D., Falter, Z., & Bhunia, S. (2023, May). A review of colonial pipeline ransomware attack. In 2023 IEEE/ACM 23rd International Symposium on Cluster, Cloud and Internet Computing Workshops (CCGridW) (pp. 8-15). IEEE.*

- Li, C. (2022). Securing US Critical Infrastructure against Cyber Attacks. In Harvard model congress.

- Smith, D. C. (2021). Cybersecurity in the energy sector: are we really prepared?. Journal of Energy & Natural Resources Law, 39(3), 265-270.

- Reeder, J. R., & Hall, T. (2021). Cybersecurity's pearl harbor moment. The Cyber Defense Review, 6(3), 15-40.

- K.,Greg,Eckels, S., (2022). How SolarWinds still affects supply chain threats two years later. Google Cloud: https://cloud.google.com/blog/products/identity-security/how-solarwinds-still-affects-supply-chain-threats-two-years-later

- Mandiant. (2020). Evasive attacker leverages SolarWinds supply chain compromises with SUNBURST backdoor: https://www.mandiant.com/resources/blog/evasive-attacker-leverages-solarwinds-supply-chain-compromises-with-sunburst-backdoor

- Martínez, J., & Durán, J. M. (2021). Software supply chain attacks, a threat to global cybersecurity: SolarWinds' case study. International Journal of Safety and Security Engineering, 11(5), 537-545

- Lazarovitz, L. (2021). Deconstructing the solarwinds breach. Computer Fraud & Security, 2021(6), 17-19.

- Case, D. U. (2016). Analysis of the cyber attack on the Ukrainian power grid. Electricity Information Sharing and Analysis Center (E-ISAC), 388(1-29), 3.

- Huang, B., Majidi, M., & Baldick, R. (2018, August). Case study of power system cyber attack using cascading outage analysis model. In 2018 IEEE Power & Energy Society General Meeting (PESGM) (pp. 1-5). IEEE.

- Baezner, M. (2018). Cyber and Information warfare in the Ukrainian conflict (No. 1, pp. 1-56). ETH Zurich.

- Perlroth, N., Scott, M., & Frenkel, S. (2017). A Cyberattack Hits Ukraine, Then Spreads. The New York Times, A1-L.

- Kao, D. Y., & Hsiao, S. C. (2018, February). The dynamic analysis of WannaCry ransomware. In 2018 20th International conference on advanced communication technology (ICACT) (pp. 159-166). IEEE.

- Mohurle, S., & Patil, M. (2017). A brief study of wannacry threat: Ransomware attack 2017. International journal of advanced research in computer science, 8(5), 1938-1940.

- Zimba, A., Wang, Z., & Chen, H. (2018). Multi-stage crypto ransomware attacks: A new emerging cyber threat to critical infrastructure and industrial control systems. Ict Express, 4(1), 14-18.

- Woollaston-Webber, V. (2017). WannaCry ransomware: what is it and how to protect yourself

- Wired: https://www.wired.com/story/wannacry-ransomware-virus-patch/

- *Goodin, D. (2017). Mysterious Microsoft patch killed 0-days released by NSA-leaking Shadow Brokers*

- *Ars Technica:* https://arstechnica.com/information-technology/2017/04/purported-shadow-brokers-0days-were-in-fact-killed-by-mysterious-patch/

- *Fayi, S. Y. A. (2018). What Petya/NotPetya ransomware is and what its remediations are. In Information technology-new generations: 15th international conference on information technology (pp. 93-100). Springer International Publishing.*

- *Watson, F. C., CISM, C., & ECSA, A. (2017). Petya/NotPetya why it is nastier than WannaCry and why we should care. ISACA, 6, 1-6.*

- *Branquinho, M. A. (2018). Ransomware in industrial control systems. What comes after Wannacry and Petya global attacks?. WIT Trans. Built Environ, 174, 329-334.*

- *Greenstein, S. (2019). The aftermath of the dyn DDOS attack. IEEE Micro, 39(4), 66-68.*

- *Scott Sr, J., & Summit, W. (2016). Rise of the machines: The dyn attack was just a practice run december 2016. Institute for Critical Infrastructure Technology, Washington, DC, USA, 3, 9.*

- *Magaña, J., Olvera, C. I., & Lous, P. (2019). How can we improve security against DDoS attacks? A case study: The DyN Attack in 2016.*

- *Kambourakis, G., Kolias, C., & Stavrou, A. (2017, October). The mirai botnet and the iot zombie armies. In MILCOM 2017-2017 IEEE military communications conference (MILCOM) (pp. 267-272). IEEE.*

- *Margolis, J., Oh, T. T., Jadhav, S., Kim, Y. H., & Kim, J. N. (2017, July). An in-depth analysis of the mirai botnet. In 2017 International Conference on Software Security and Assurance (ICSSA) (pp. 6-12). IEEE.*

Part 3: Protecting Critical Infrastructure

Part 3 outlines a comprehensive strategy for cyber defense, beginning with network security, advancing through continuous monitoring, and then progressing to the implementation of robust security policies and frameworks. This section progresses from technical safeguards to fostering a culture of security awareness and readiness for incident response, also considering the impact of executive orders on cybersecurity practices. It integrates these elements to form a full-spectrum defense approach for critical infrastructure protection.

This part has the following chapters:

Protecting Critical Infrastructure – Part 1

Welcome to *Chapter 6* of our exciting journey through the world of cybersecurity! In this installment, we'll explore more of the vital topic of protecting critical infrastructure. Over the next pages, we will explore the essential cybersecurity techniques crucial in the battle against cyber threats aimed at our critical infrastructure.

Imagine a world where the systems that power our cities, transport, healthcare, and economy are under constant threat from cyberattacks. Well, that world exists today, and it's our responsibility to safeguard these critical assets. This chapter is your guide to understanding the intricacies of fortifying our critical infrastructure against digital adversaries.

In the preceding chapter, we explored the intricate world of cyber conflict, where the digital and physical realms intertwine, often with profound consequences. We witnessed the alarming power of cyberattacks, from Stuxnet's eerie silence to the ransom waves of WannaCry and NotPetya, and the vulnerabilities exposed in the SolarWinds incident. The breaches affecting essential infrastructure, such as the Colonial Pipeline, left an indelible mark on our collective awareness.

As we stand on the precipice of an increasingly interconnected future, the need for robust cyber defenses has never been more apparent. In this chapter, we shift our focus from the dark tales of cyber warfare to the strategies and solutions that can safeguard our digital existence. We will explore a comprehensive range of mitigations and defenses, from technical measures to organizational practices, that can help fortify our systems against the looming threats.

Join us as we embark on a journey to navigate the complex labyrinth of cybersecurity, where knowledge is the ultimate shield, and preparedness is our greatest armor. Together, we will uncover the key principles and practices necessary to secure our interconnected world and protect the very foundation of our digital age.

In this chapter, we will cover the following topics:

- Network security and continuous monitoring
- Security policy and frameworks

Network security and continuous monitoring

Network security and security monitoring are two interrelated aspects of information technology and cybersecurity that are crucial for safeguarding computer networks and systems from a wide range of threats, including unauthorized access, data breaches, malware, and other cyberattacks that are of critical importance to critical infrastructure.

Network security refers to the set of practices, measures, and technologies put in place to protect the confidentiality, integrity, and availability of computer networks and their associated resources, including data, applications, and devices. Its primary goal is to ensure that only authorized users and devices have access to network resources while preventing malicious actors from compromising the network.

Let's look at some of the essential techniques and practices for enhancing network security.

Network segmentation

Segment your network to isolate critical systems from less critical ones. This can help contain lateral movement and the spread of malware if it infiltrates your network.

Network segmentation is a vital security strategy that entails partitioning a network into smaller, isolated sections or zones. This is done to minimize the exposed attack area and restrict the ability of attackers to move laterally within the network. The following table shows some basic techniques for network segmentation to enhance security:

Technique	Description
Subnetting	Divide your network into subnets, each with its own IP address range. This provides a foundational level of segmentation and helps isolate different departments or functions within your organization.
Virtual local area networks (VLANs)	VLANs allow you to create logical network segments within a physical network infrastructure. Devices within the same VLAN can communicate with each other, but traffic between VLANs is restricted by default. This provides isolation and control.
Firewalls	Place firewalls at strategic points in your network to control traffic between segments. You can create rules and policies to allow or deny specific types of traffic between segments, enhancing security.

Technique	Description
Access control lists (ACLs)	ACLs are used in routers and switches to define what traffic is allowed or denied between different network segments. They can be a part of firewall rules or used independently to control traffic.
Network address translation (NAT)	NAT can be used to hide internal network addresses and provide an additional layer of security. It can be employed at the boundary between segments to control which IP addresses are exposed externally.
Port address translation (PAT)	PAT allows multiple devices on a local network to be mapped to a single public IP address but with a different port number for each session. This is useful for saving IP addresses and for increasing security, as it hides the individual IP addresses of internal network devices from external networks.
Proxy servers	Implement proxy servers to manage and inspect traffic between network segments. Proxies can act as intermediaries, providing an additional layer of security and control over communication.

Table 6.1 – Network segmentation

Let's illustrate the preceding with a hypothetical scenario for a power grid substation network segmentation strategy, where the network is divided into different security zones based on the criticality and sensitivity of the devices and systems.

Control zone

This is the most critical zone containing the **supervisory control and data acquisition (SCADA)** systems and critical infrastructure control systems. Access to this zone should be highly restricted and monitored.

Perimeter zone

This zone surrounds the control zone and contains firewalls, intrusion detection systems, and security gateways. It acts as a buffer between the control zone and the less critical zones.

Security monitoring zone

This zone contains security systems for monitoring and logging network traffic. **Security information and event management (SIEM)** systems are often located here to analyze logs for potential threats.

Enterprise zone

This is where the administrative and business systems are located, such as email, HR, and other non-critical applications. It is isolated from the control and monitoring zones.

Demilitarized zone

The **demilitarized zone (DMZ)** is a neutral zone between the internal network and the external internet. It contains servers and services that need to be accessible from the internet, such as a public-facing website or remote monitoring systems. The DMZ acts as a barrier to prevent direct access to the control zone.

Let's continue with another critical component of the network security strategy: access control.

Access control

Access control is a fundamental aspect of network security that involves regulating who can access specific resources, devices, or areas within a network. Effective access control techniques help prevent unauthorized users or devices from gaining access to sensitive data or systems.

Implement strict access controls to limit who can access the network, systems, and data. Use strong authentication methods and enforce the principle of least privilege (granting the minimum level of access needed for users or devices).

Here are some common techniques for access control in network security:

Technique		Description
User authentication	Username and password	This is the most common method, where users must enter a username and password to access a network resource.
	Multi-factor authentication (MFA)	Requires users to provide at least two forms of authentication, typically something they know (password) and something they have (e.g., a smartphone app or hardware token).
	One-time passwords (OTPs)	Commonly used as part of MFA, OTPs are generated using a cryptographic key, a seed value, and a time-based algorithm. The key is shared between the authentication server and the user's device. Each OTP is unique and only valid for a short time, providing strong authentication.
	Biometric authentication	Uses unique biological traits such as fingerprints, retinal scans, or facial recognition for user authentication.
	SSH keys and certificates	SSH keys and certificates are used for secure access to network devices and servers, replacing traditional username/password authentication.

Technique	Description	
ACLs	**Network devices**	Routers and switches can use ACLs to define rules that permit or deny access based on IP addresses, ports, or protocols.
	Firewalls	Firewall rules specify which traffic is allowed or blocked based on source and destination IP addresses, ports, and application-layer data.
Role-based access control (RBAC)	RBAC assigns users and devices to specific roles with predefined permissions. Users inherit access rights associated with their roles.	
Attribute-based access control (ABAC)	Access is granted or denied based on various attributes such as user attributes (e.g., department, job title, etc.), resource attributes (e.g., sensitivity level), and environmental attributes (e.g., time of day).	
Mandatory access control (MAC)	MAC enforces access controls based on labels or security clearances and is typically used in government or high-security environments.	
Discretionary access control (DAC)	DAC allows resource owners to define access permissions for their resources.	
Access tokens	Access tokens can be issued to users after authentication, containing information about their access rights.	
Session management	Session management ensures that users are authenticated only once and can access resources without repeatedly entering credentials during a session.	
Guest networks	Separate networks are created for guest or untrusted users, limiting their access to the organization's internal resources.	
Directory services (e.g., Active Directory)	Centralized directory services enable the management of user accounts, group memberships, and access controls across the network.	
Dynamic access control	Dynamic access control policies can automatically assign access permissions based on file attributes, user attributes, and conditions.	
Network access control (NAC)	NAC solutions enforce security policies for devices seeking access to the network, ensuring that only compliant and trusted devices can connect.	
API access controls	Implement access controls for APIs to restrict who can interact with your applications and services programmatically.	

Table 6.2 – Access control in network security

The choice of access control techniques and the level of security they provide should align with the organization's security policies, risk assessment, and regulatory compliance requirements. Access control is a critical component of network security and plays a significant role in protecting sensitive data and resources from unauthorized access.

Intrusion detection and prevention systems

Intrusion detection systems (IDS) and **intrusion prevention systems (IPS)** are critical components of network security. They play a significant role in identifying and responding to security threats and anomalies in real time.

Place IDS/IPS solutions between network segments to monitor and filter traffic for malicious activity. They can help prevent attacks from penetrating and spreading across the network.

Here's an overview of IDS:

IDS		
Purpose	An IDS is a security solution designed to monitor network traffic and system activity to identify and alert on suspicious or malicious behavior. It acts as a watchdog for your network.	
Type	**Network-based IDS (NIDS)**	A NIDS monitors network traffic and looks for suspicious patterns or signatures. It can be placed strategically within the network to inspect all traffic.
	Host-based IDS (HIDS)	A HIDS focuses on individual devices (hosts) and inspects activities on the device, such as file changes, login attempts, and system calls.
Detection methods	**Signature-based detection**	This compares observed network traffic or system activity to a database of known attack signatures. It's effective for detecting known threats but may miss new, unknown attacks.
	Anomaly-based detection	This establishes a baseline of normal network or system behavior and triggers alerts when deviations from this baseline occur. It can identify previously unknown threats but may produce false positives.
	Hybrid detection	This combines signature-based and anomaly-based detection methods for more comprehensive threat detection.
Alerts and reporting	When an IDS detects suspicious activity, it generates alerts and logs the relevant information. Security administrators analyze these alerts to determine whether they indicate a security threat.	
Passive monitoring	An IDS typically operates in a passive monitoring mode. It observes and alerts on suspicious activity but does not actively block or prevent attacks.	

Table 6.3 – Overview of IDSs

The following table gives an overview of IPS:

IPS		
Purpose	IPS builds upon IDS capabilities by not only detecting threats but also taking automated actions to prevent or mitigate them in real time. It acts as a security enforcer.	
Type	**Network-based IPS (NIPS)**	A NIPS monitors network traffic and, in addition to detecting threats, actively blocks or drops malicious traffic to prevent attacks.
	Host-based IPS (HIPS)	A HIPS protects individual devices by inspecting system activity and taking actions such as blocking specific applications or system functions.
Signature-based and anomaly-based prevention	IPSs can use both signature-based and anomaly-based techniques for threat prevention. Signature-based prevention is effective against known threats, while anomaly-based prevention helps defend against novel attacks.	
Detection and response	The IPS can take various actions in response to detected threats, such as blocking traffic from specific IP addresses, dropping malicious packets, or resetting connections. The goal is to actively prevent or mitigate attacks.	
In-line deployment	IPSs are often deployed in line with network traffic, meaning that traffic must pass through the IPS before reaching its destination. This enables real-time threat prevention.	
Out-of-band or passive deployment	IPS systems monitor a copy of network traffic and can provide post-analysis and reporting without directly affecting the traffic flow.	
False positives and tuning	Fine-tuning an IPS is crucial to reduce false positives, which can disrupt legitimate traffic. A balance must be struck between security and operational efficiency.	

Table 6.4 – Overview of IPSs

Both IDS and IPS are valuable components of a network security strategy. IDS helps identify potential threats and provides insights for further investigation, while IPS goes a step further by actively preventing or mitigating those threats. The choice between IDS and IPS depends on your security requirements, operational needs, and risk tolerance. Many organizations deploy a combination of both to achieve a comprehensive security posture.

IDS and IPS are especially critical for safeguarding the security of critical infrastructure. Here are a few examples of how IDS and IPS can be applied in critical infrastructure protection.

Energy grid security

A utility company responsible for an electrical grid deploys IDS/IPS systems to protect critical infrastructure components. The systems can be integrated into substations, control centers, and remote access points to monitor network traffic and system activities.

These security solutions continuously analyze traffic patterns and look for suspicious activities, particularly those that could indicate a cyberattack.

The IDS detects an anomalous pattern in network traffic, suggesting that an attacker is attempting to gain unauthorized access to the grid's control system and generates an alert, and the security team is notified of the potential intrusion attempt.

The security team investigates the incident, and the IPS takes immediate action to block the attacker's IP address, preventing any further unauthorized access attempts. Simultaneously, network segments may be isolated to limit damage and protect critical infrastructure.

Transportation and traffic management

A city's transportation authority uses IDSs/IPSs to secure traffic control systems and ensure the integrity of transportation infrastructure.

The IDS/IPS can be installed in traffic management centers, controlling systems for traffic lights, and electronic road signs. These systems monitor communication between various traffic management components and detect any unauthorized access or manipulation. An IDS can detect an unauthorized user attempting to manipulate traffic signal patterns, potentially causing traffic chaos. Potentially, alerts are generated and sent to the security team, notifying them of the intrusion attempt. The IPS immediately blocks the unauthorized user, and the security team works to trace the origin of the attack, investigate vulnerabilities, and strengthen access controls to prevent future attacks on the transportation infrastructure.

Water treatment facilities

A water treatment facility relies on an IDS/IPS to secure its critical systems and ensure the safety of the water supply.

IDSs/IPSs are deployed in the control room, monitoring SCADA systems, and remote access points. These systems monitor network traffic, looking for anomalies that could indicate unauthorized access or malicious manipulation of water treatment systems.

The IDS could detect unusual patterns in network traffic, potentially indicating an unauthorized user attempting to alter water treatment processes. An alert about this activity can be generated by the IDS and sent to the facility's security team, who then initiates an incident response if necessary.

The IPS can take immediate action to block the unauthorized user, while the security team investigates the breach and reinforces the facility's cybersecurity measures to protect the critical infrastructure.

In these examples, IDSs/IPSs play a crucial role in securing critical infrastructure against cyber threats and helping ensure the continued operation and safety of essential services, including energy grids, transportation systems, and water treatment facilities. It highlights the importance of implementing comprehensive security measures to protect critical infrastructure from potential attacks.

Virtual private networks (VPNs)

VPNs play a significant role in securing critical infrastructure by providing a secure and encrypted communication channel for remote access, monitoring, and management of critical systems and data. Here's how VPNs are used in critical infrastructure protection:

Component	Description
Secure remote access	VPNs enable authorized personnel to securely access critical infrastructure systems and data remotely. This is crucial for troubleshooting, maintenance, and monitoring without the need for physical presence at the infrastructure sites.
Protection of data in transit	VPNs encrypt data in transit, ensuring that sensitive information exchanged between remote users and the critical infrastructure remains confidential and secure. This is essential for maintaining data integrity.
Protection of communication between facilities	Critical infrastructure often involves multiple facilities and locations. VPNs secure communication between these sites, safeguarding the data and control signals exchanged between them.
Access control and authentication	VPNs often incorporate strong user authentication and access control mechanisms, ensuring that only authorized personnel can establish connections to critical infrastructure systems. VPNs may support MFA, requiring users to provide multiple forms of authentication before gaining access to the network.
Redundancy and failover	VPN solutions can be configured for redundancy and failover to ensure continuous and reliable access to critical infrastructure systems. This minimizes downtime and ensures operations remain uninterrupted.
Tunneling protocols	VPNs use tunneling protocols to encapsulate and secure data traffic. Common protocols include **Internet Protocol Security (IPsec)** and **Layer 2 Tunneling Protocol (L2TP)**.
Security auditing and monitoring	VPN solutions often include logging and auditing capabilities, allowing organizations to monitor who is accessing the network and detect any unusual or suspicious activities.

Component	Description
Data center and cloud connectivity	VPNs can securely connect critical infrastructure components to data centers or cloud services, providing scalability and flexibility while maintaining a high level of security.
Compliance and regulatory requirements	Many critical infrastructure sectors are subject to strict compliance and regulatory standards. VPNs help organizations meet these requirements by ensuring secure communications and data protection.

Table 6.5 – Usage of VPNs in CI protection

In the context of critical infrastructure protection, VPNs are just one component of a comprehensive security strategy. It's important to combine VPN technology with other security measures, such as IDSs/IPSs, firewall protection, network segmentation, and robust access controls, to create a layered defense system that safeguards these vital assets from cyber threats.

Security audits and penetration testing

Securing critical infrastructure through comprehensive security audits and penetration testing involves a multi-faceted approach. Various techniques are employed to identify vulnerabilities, assess risks, and fortify systems against potential cyber threats. The following sections outline some key techniques utilized for safeguarding critical infrastructure.

Vulnerability assessment

Vulnerability assessment involves systematically identifying, quantifying, and prioritizing vulnerabilities within the infrastructure. This process often includes automated tools to scan networks, systems, and applications for weaknesses. These assessments help in understanding potential entry points for attackers and provide a baseline for subsequent security measures.

Network scanning and mapping

Network scanning tools are utilized to comprehensively map the infrastructure, identifying all connected devices, open ports, and potential vulnerabilities. This process aids in understanding the network topology and potential weak points that might be exploited by intruders.

Social engineering assessments

As humans are often the weakest link in cybersecurity, social engineering assessments simulate real-world scenarios where attackers attempt to manipulate individuals to divulge sensitive information or grant unauthorized access. These assessments help in training and preparing staff against social engineering tactics.

Penetration testing

Penetration testing, also known as **pen testing**, involves simulating cyberattacks to assess the security posture of critical infrastructure. Ethical hackers, security professionals, or specialized teams attempt to breach the system using techniques mirroring those of real attackers. This process uncovers vulnerabilities and provides insights into the effectiveness of existing security measures.

Red team/blue team exercises

In this technique, a *red team* acts as the attacker, attempting to breach the infrastructure's security, while the *blue team* defends against these attacks. These exercises provide real-time simulations, enabling teams to practice their response to potential active threats and improve security protocols.

Log analysis and monitoring

Analyzing logs and monitoring network activities in real time is critical for detecting and responding to potential security incidents. Log analysis tools help in identifying anomalies, suspicious activities, and potential breaches, enabling swift responses to mitigate risks.

Compliance audits

Conducting compliance audits ensures that critical infrastructure adheres to industry regulations, standards, and best practices. This helps in maintaining a baseline level of security and ensuring that the infrastructure meets required security standards.

Implementing these techniques in a coordinated and systematic manner is essential for protecting critical infrastructure. Regular and thorough security audits and penetration testing help in identifying weaknesses, fortifying defenses, and maintaining the resilience of these vital systems against ever-evolving cyber threats.

Honeypots and deception technologies

Honeypot and deception techniques are valuable strategies in the realm of cybersecurity, especially for safeguarding critical infrastructure. These methods involve creating decoy systems, data, or networks to divert and deceive potential attackers, gather information about their tactics, and protect the actual infrastructure.

Honeypots are decoy systems or resources deliberately designed to attract attackers. There are two main different types of honeypots:

- **Research honeypots** are designed to gather information about attackers' methods and tools. They're typically placed in public-facing areas of a network and/or unused IP space to lure and observe attackers without impacting critical systems.

- **Production honeypots** are integrated within the operational network to deceive and divert attackers from actual critical systems. They're used as an early warning system, triggering alerts when unauthorized access is attempted.

Honeypots can detect attacks in their early stages, allowing security teams to respond swiftly.

They provide valuable insights into attackers' methods, tools, and strategies, which can be used to strengthen security measures. By drawing attackers away from critical infrastructure, they act as a form of defense.

Deception techniques involve creating a false environment or information to mislead attackers and steer them away from actual critical systems; for example, placing false or misleading files and data within systems to confuse attackers and lead them astray. Manipulate network traffic to deceive attackers about the structure and nature of the network, making it harder for them to identify and access real critical assets, and create fake credentials or access points that, if accessed, trigger alerts and help identify potential threats.

Deception techniques make it challenging for attackers to distinguish actual critical assets from fake ones.

Attackers spend their time and resources on false information or systems, slowing down their progress toward critical infrastructure. Alerts are triggered when the deceptive elements are accessed, allowing for proactive security measures.

Both honeypots and deception techniques are powerful tools, but they require careful planning and monitoring. Regular monitoring and maintenance are essential to ensure the effectiveness of these techniques. While honeypots and deceptions divert attackers, there's a slight risk that they might not be as effective against sophisticated attackers who can recognize these decoys.

Integrating these techniques within a comprehensive security strategy for critical infrastructure can significantly enhance defense capabilities and provide crucial insights into potential active threats, ultimately bolstering the overall cybersecurity posture.

Zero trust architecture

Zero trust architecture (ZTA) is a security concept and approach that assumes no implicit trust within an organization's network, whether it's inside or outside the perimeter. For critical infrastructure, ZTA plays a crucial role in fortifying defenses and mitigating potential risks. Here's why it's vital:

- **Limits lateral movement**: In critical infrastructure, once an attacker gains access, the ability to move laterally within the network poses a significant threat. ZTA enforces strict access controls, requiring verification for every access attempt. This limits an attacker's ability to move freely within the network, reducing the impact of a breach.

- **Enforces continuous verification**: ZTA operates on the principle of continuous verification and authentication. It doesn't assume that once someone gains access, they are inherently trustworthy. Each access request is thoroughly validated, regardless of the user's location or the network they're trying to access. This minimizes the chances of unauthorized access, even if an attacker gains a foothold within the system.

- **Protects critical assets**: For critical infrastructure, protecting the most vital assets is key. ZTA allows for granular control and segmentation of these assets, ensuring that only authorized and authenticated users can access them. Even if other parts of the network are compromised, this approach provides an additional layer of defense for the most critical components.

- **Adapts to change**: ZTA is adaptable to evolving threats and technologies. As critical infrastructure systems evolve, ZTA can be updated and integrated with new security measures, ensuring that security protocols remain up to date.

- **Reduces attack radius**: In the unfortunate event of a breach, ZTA limits the extent of damage. By compartmentalizing access and implementing strict controls, the impact of a breach is confined to the specific area compromised, reducing the overall damage and allowing for more effective containment and response.

In critical infrastructure, where the stakes are high and the potential consequences of a security breach are significant, implementing ZTA is a proactive and strategic approach. It challenges the traditional perimeter-based security model, providing a more comprehensive and dynamic defense strategy that continually validates and verifies every access attempt, ultimately enhancing the overall security posture of critical systems.

Security monitoring

Security monitoring involves the continuous and systematic observation of a network or system to identify and respond to security events and incidents. The key objective of security monitoring is to detect anomalous or suspicious activities that may indicate a security breach or violation of security policies.

For critical infrastructure, early threat detection is crucial. Network security monitoring uses a range of tools and techniques to detect anomalies, unusual activities, or potential security breaches. Identifying these issues in their nascent stages allows for timely intervention and mitigation, minimizing potential damage.

To establish a robust network security monitoring system, it is imperative to meticulously design a comprehensive security stack. This security framework assumes the responsibility of filtering malicious traffic, preempting the emergence of compromised systems, and streamlining incident response.

Crafting an effective security stack entails a structured amalgamation of various security layers, each serving a distinct yet integral function in fortifying network defenses. These layers collectively contribute to the following key objectives:

- Firstly, the security stack operates as a filtering mechanism, discerning and obstructing malicious traffic attempting to breach the network's perimeter. This initial defense line involves tools such as firewalls, IPSs, and secure gateways, aiming to restrict unauthorized access and identify potentially harmful packets.

- Secondly, it endeavors to avert the compromise of systems within the network. Endpoint protection tools, encompassing antivirus software, endpoint detection and response solutions, and comprehensive device management systems, play a critical role in shielding individual devices from malware, unauthorized access, and other potential threats.

- Thirdly, the security stack is pivotal in facilitating incident response. This involves the implementation of monitoring tools, IDSs, and SIEM solutions to detect anomalies and suspicious activity. These tools contribute significantly to the timely identification, containment, and resolution of potential security incidents.

By orchestrating a meticulously designed security stack, organizations can fortify their network security monitoring capabilities. This amalgamation of diverse security measures ensures a proactive defense against a constantly evolving spectrum of cyber threats, securing the integrity and resilience of critical network infrastructures.

Optical fiber taps

Optical fiber taps are devices used to access and intercept data transmitted through fiber optic cables without disrupting the flow of information. These taps are designed to discreetly monitor and capture the data passing through the fiber optic lines.

Optical fiber taps operate by diverting a small portion of the light signal traveling through the fiber optic cable. They use a technology that allows for minimal intrusion into the cable, ensuring that most of the data continues its journey uninterrupted. This diverted portion is then directed to a security stack and monitoring devices for analysis or interception purposes.

Here are some types of taps:

Type	Description
In-line taps	These taps are installed directly into the fiber optic cable, splitting the light signal and siphoning off a portion for monitoring.
Splitter taps	A passive optical splitter is placed on the fiber line to split the signal, allowing a portion of the light to be directed to the tap for monitoring purposes.
Reflective taps	These taps use the principle of reflecting a small portion of the light signal from the fiber, diverting it to the monitoring system for analysis.

Table 6.6 – Types of optical taps

Taps are utilized by network administrators for monitoring network traffic and analyzing data for performance or security purposes. Law enforcement agencies or security entities may use fiber taps for lawful interception or surveillance purposes. They also aid in diagnosing issues, testing the network's performance, and troubleshooting potential problems in the transmission of data.

Optical fiber taps play a significant role in network monitoring, security, and troubleshooting within the realm of fiber optic communication systems, providing a method for discreetly accessing and analyzing data without interrupting the flow of information.

Network packet brokers (NPBs)

NPBs are critical components in modern network infrastructures. They play a central role in security monitoring by intelligently managing and distributing network traffic for analysis and security tools.

NPBs collect and aggregate network traffic from multiple sources, including switches, routers, and other network devices. They filter and organize this traffic, ensuring that security tools receive only the relevant data they require for analysis.

These devices optimize the traffic flow by directing packets to appropriate security and monitoring tools, ensuring that these tools operate at their maximum efficiency and effectiveness.

NPBs distribute traffic evenly across multiple security tools, load balancing and preventing overloading of any single tool and maintaining consistent performance across the security infrastructure. They can manipulate packets by slicing them (dividing them into smaller sections) or removing duplicate packets to enhance the efficiency of analysis tools and reduce unnecessary data processing.

NPBs can encapsulate and encrypt data, ensuring that sensitive information is protected during transmission to security tools for analysis.

NPBs play a pivotal role in the efficient and effective functioning of security monitoring infrastructure. By optimizing network traffic, they enable security tools to focus on relevant data, enhancing the overall security posture and responsiveness of an organization's network defenses.

Next-generation firewalls

Next-generation firewalls (**NGFWs**) represent a significant evolution in network security. These advanced systems go beyond traditional firewall functionalities, integrating additional features for enhanced network monitoring, threat detection, and mitigation.

Here are some key capabilities of next-generation firewalls used for network security monitoring:

Capability	Description
Deep packet inspection (DPI)	NGFWs conduct thorough inspections of network packets, not just based on port or protocol, but on the application layer, enabling granular visibility into the data being transmitted. This allows for better control and monitoring of applications and associated risks.
Application visibility and control	NGFWs provide detailed insights into network traffic, allowing administrators to monitor and control the applications being used on the network. This helps in identifying potential security risks associated with specific applications.
IPS	Advanced IPS capabilities are integrated into NGFWs, enabling real-time detection and prevention of potential threats. This includes the ability to identify and block known malware, attack patterns, and other malicious activities.
User identity awareness	NGFWs can associate network traffic with specific users, allowing for personalized security policies and monitoring of individual user activities.
Threat intelligence integration	NGFWs often incorporate threat intelligence feeds, enabling them to proactively block traffic associated with known malicious sources or activities.
SSL inspection	NGFWs can inspect encrypted traffic (SSL/TLS), decrypting and analyzing the content within these encrypted connections to identify potential threats.
Behavioral analytics	Some NGFWs incorporate behavioral analytics, allowing them to detect anomalies in network traffic behavior, which could indicate potential security risks.

Table 6.7 – Key capabilities of NGFWs

NGFWs play a crucial role in network monitoring by providing advanced capabilities that go beyond traditional firewall functionalities. These devices offer comprehensive visibility and control over network traffic, significantly enhancing an organization's ability to monitor, detect, and respond to potential security threats.

IDS

As described in depth earlier in this chapter, IDSs are crucial in a security stack for security monitoring purposes. Specifically, IDSs play a crucial role in early threat detection, providing real-time alerts when suspicious activities or potential security breaches are detected. They offer enhanced visibility into network and host activities, providing security analysts with insights into potential threats and vulnerabilities. IDS alerts enable swift incident response, allowing security teams to investigate and mitigate potential security incidents in a timely manner.

Log analysis tools

Log analysis tools are crucial in the realm of security monitoring, enabling organizations to collect, analyze, and derive insights from various logs generated by network devices, applications, and systems. These tools play a significant role in identifying anomalies, detecting potential security threats, and facilitating incident response. Here's a breakdown of their functions and key features:

Capability	Description
Log collection	Log analysis tools gather logs generated by a wide range of devices and applications, including servers, network equipment, firewalls, and security systems such as IDS.
Aggregation and centralization	These tools centralize log data, aggregating it into a single location, making it easier to analyze and manage.
Normalization and parsing	Log analysis tools normalize and parse logs, organizing the data into a consistent format for easier analysis and correlation.
Correlation and analysis	They analyze logs in real time, correlating data from various sources to identify patterns, anomalies, and potential security threats.
Alerting and reporting	These tools generate alerts and reports based on predefined rules and thresholds, allowing security teams to act promptly in response to potential security incidents.
Search and query capabilities	These tools provide powerful search and query functionalities, enabling security analysts to search through vast amounts of log data efficiently.
Customizable dashboards	They offer customizable dashboards and visual representations, allowing for easy visualization of key metrics and trends within the log data.
Anomaly detection	Some log analysis tools use machine learning or pattern recognition to identify abnormal behavior or deviations from normal patterns.

Capability	Description
Compliance and forensics support	Many tools offer functionalities to aid in compliance adherence and forensic investigations by providing historical log data and audit trails.
Integration and scalability	They often integrate with other security tools and systems and can scale to accommodate large volumes of log data.

Table 6.8 – Capabilities of log analysis tools

Log analysis tools play a pivotal role in detecting security incidents by identifying unusual patterns or activities that may indicate a security threat. They facilitate efficient incident response by providing real-time alerts and reports that help security teams investigate and mitigate potential security breaches.

In summary, log analysis tools are indispensable for security monitoring, providing the means to collect, analyze, and act upon the vast amount of log data generated within an organization's network, thereby strengthening security postures and ensuring proactive threat detection and response.

Black hole routing

Black hole routing refers to a network management technique where incoming traffic destined for a specific destination is dropped, essentially creating a *black hole* where the data is discarded without being forwarded. This technique is often used in network security or response to certain network issues.

The routing capability involves directing traffic to a null or non-existent location, which effectively discards the packets. It can be used deliberately for security purposes, such as mitigating certain types of attacks such as a DDoS attack, where traffic from suspicious sources or targeting specific vulnerabilities is sent to the black hole instead of reaching the intended destination.

Black hole routing can also be accidental, resulting from misconfigurations or network failures, where traffic intended for a valid destination is mistakenly routed to a non-existent or unreachable destination, effectively causing a loss of data.

Network administrators and security professionals use black hole routing selectively and with caution, as it can impact legitimate traffic if not implemented correctly. The technique is also often employed temporarily as a reactive measure to mitigate threats or address network issues until a more permanent solution is established.

In summary, network security encompasses the practices and technologies that protect a network's integrity and resources, while security monitoring involves the continuous observation of network and system activities to detect and respond to potential security threats and incidents. Both aspects are critical in maintaining a strong cybersecurity posture and mitigating the risks associated with the ever-evolving threat landscape.

Security policy and frameworks

Securing critical infrastructure is a complex task that requires the implementation of various security policies and frameworks to mitigate risks and protect against cyber threats. The following subsections outline some key security policies and frameworks that can be applied to critical infrastructure.

NIST cybersecurity framework

Developed by the **National Institute of Standards and Technology (NIST)**, this framework provides a structured approach to managing and reducing cybersecurity risk. It consists of five core functions: *identify, protect, detect, respond,* and *recover,* which can be tailored to address the unique needs of critical infrastructure sectors.

Most specifically, NIST Special Publication 800-53, titled *Security and Privacy Controls for Information Systems and Organizations,* is a publication by the NIST in the United States. It provides a comprehensive set of security controls and guidelines for federal information systems and organizations to enhance the security and privacy of their information systems.

NIST Special Publication 800-53 is a key document within the broader NIST Special Publication 800 series, which covers various aspects of information security, including risk management, cybersecurity, and privacy. Special Publication 800-53 is particularly important for federal agencies and organizations that handle sensitive and classified information, as it helps them establish a robust security and privacy framework.

The document outlines a catalog of security controls that can be customized and tailored to meet the specific security needs and requirements of different federal information systems. These controls cover a wide range of security areas, including access control, authentication, audit and accountability, incident response, and encryption.

The following table shows the key components of NIST Special Publication 800-53:

Component	Description
Control families	The controls are organized into 20 control families, each addressing a specific aspect of security or privacy. Examples of control families include access control, system and communication protection, security assessment and authorization, and privacy.
Control baselines	NIST Special Publication 800-53 provides various security control baselines that are tailored for different types of federal information systems, such as low-impact, moderate-impact, and high-impact systems. These baselines help organizations apply the appropriate level of security controls based on their system's sensitivity and risk.

Component	Description
Control descriptions	Each control is accompanied by a detailed description, implementation guidance, and references to related security standards and guidelines.
Security control selection	Organizations are expected to select and tailor the security controls from the catalog to align with their specific security requirements and operational environment.
Continuous monitoring	The publication emphasizes the importance of continuous monitoring of security controls and systems to detect and respond to security incidents and vulnerabilities effectively.

Table 6.9 – NIST Special Publication 800-53 key components

NIST Special Publication 800-53 is regularly updated to reflect changes in technology, threats, and best practices in information security. Organizations that adhere to these guidelines are better equipped to protect their information systems and sensitive data from cyber threats and security breaches. It also serves as a valuable resource for organizations outside the federal sector seeking robust cybersecurity and privacy controls.

ISO/IEC 27001 and ISO/IEC 27002

The **ISO 27001** standard provides a systematic approach to managing and protecting information assets. It covers risk assessment, security policies and procedures, access control, and incident response, making it suitable for safeguarding critical infrastructure systems and data.

ISO/IEC 27001 is an **information security management system (ISMS)** standard, which specifies the requirements for establishing, implementing, maintaining, and continually improving an ISMS within an organization. It provides a framework for organizations to manage information security and risks systematically.

ISO/IEC 27002, however, is a code of practice for information security controls. It offers detailed guidelines and best practices for implementing security controls that organizations should consider when implementing their ISMS based on ISO 27001.

NERC CIP

The **North American Electric Reliability Corporation (NERC) critical infrastructure protection (CIP)** standards are specifically designed for the energy sector. They establish requirements for securing the **bulk power system (BPS)**, including access controls, security awareness, and incident response planning.

These standards are particularly crucial for the electric utility industry, as they help protect against cyber threats and vulnerabilities that could disrupt the electrical grid and have a significant impact on society.

The following table shows some of the key components of the NERC CIP:

Component	Purpose	Technical details
CIP-002 – BES Cyber System Categorization	Identifies and categorizes bulk electric system (BES) cyber assets to determine their criticality and the associated cybersecurity requirements.	Organizations must create an inventory of BES cyber assets, categorize them into high, medium, or low impact, and establish associated security requirements.
CIP-003 – Security Management Controls	Establishes security management controls, including policies, processes, and procedures, to protect BES cyber assets.	Requires the development and documentation of cybersecurity policies and procedures, as well as processes for risk assessment and security awareness.
CIP-004 – Personnel and Training	Ensures that personnel with access to BES cyber systems are trained and qualified to perform their job functions securely.	Specifies requirements for personnel training, background checks, and access controls for individuals with authorized cyber or physical access to critical systems.
CIP-005 – Electronic Security Perimeter	Defines the requirements for securing the **electronic security perimeter** (ESP) that protects BES cyber systems.	Requires the establishment of a security perimeter around critical systems, along with access controls, monitoring, and incident reporting.
CIP-006 – Physical Security of BES Cyber Systems	Addresses physical security measures to protect BES cyber systems from unauthorized access and tampering.	Specifies physical security requirements, including access controls, monitoring, and maintenance of physical security perimeters.
CIP-007 – Systems Security Management	Establishes requirements for managing the security of BES cyber systems, including change control and patch management.	Requires the implementation of processes for managing changes to critical systems and the identification and management of security vulnerabilities.

Component	Purpose	Technical details
CIP-008 – Incident Reporting and Response Planning	Ensures that utilities have procedures for identifying, reporting, and responding to cybersecurity incidents.	Requires the development of an incident response plan and the reporting of significant cybersecurity incidents to appropriate authorities.
CIP-009 – Recovery Plans for BES Cyber Systems	Ensures that utilities have plans to recover BES cyber systems in a security incident.	Requires the development of recovery plans for critical systems and periodic testing of these plans.
CIP-010 – Configuration Change Management and Vulnerability Assessments	Addresses the management of configuration changes and the assessment of vulnerabilities in BES cyber systems.	Requires the implementation of processes for managing and documenting configuration changes and regular vulnerability assessments.
CIP-011 – Information Protection	Focuses on protecting sensitive information related to BES cyber systems from unauthorized access and disclosure.	Requires the encryption of sensitive data, access controls, and data handling procedures.
CIP-014 – Physical Security	Establishes requirements for physical security measures to protect against physical attacks and vandalism.	Addresses physical security risk assessments, monitoring, and protection measures.

Table 6.10 – NERC CIP key components

It's important to note that the NERC CIP standards are subject to updates and revisions, and compliance requirements may change over time. Organizations subject to these standards are responsible for staying current with the latest versions and requirements to ensure the security and reliability of the electrical grid. Additionally, regulatory bodies and compliance auditors play a role in enforcing these standards and verifying compliance within the electric utility industry.

The Department of Homeland Security (DHS) critical infrastructure security framework

The US DHS provides guidelines and resources tailored to specific critical infrastructure sectors, such as transportation, healthcare, and water supply. These frameworks offer sector-specific guidance on security best practices.

Transportation Systems Sector (TSS) Cybersecurity Framework

The **TSS Cybersecurity Framework** is a set of guidelines and best practices developed by the US DHS in collaboration with the transportation sector. The framework aims to enhance the cybersecurity resilience of transportation systems, including aviation, highways, maritime, mass transit, rail, and pipelines. The following are key components and characteristics of the TSS Cybersecurity Framework:

- The TSS Cybersecurity Framework provides detailed guidance and best practices for each core function and implementation tier. It offers recommendations for safeguarding critical transportation assets, securing networks, and responding to cybersecurity incidents.

- Collaboration and information sharing are fundamental aspects of the framework. The DHS collaborates with transportation sector stakeholders, including government agencies, private industry, and associations, to ensure the framework remains relevant and effective.

- The TSS Cybersecurity Framework is designed to align with other recognized cybersecurity frameworks and standards, such as the NIST Cybersecurity Framework. This allows organizations to integrate the TSS Framework with their broader cybersecurity initiatives.

Let's explore some other DHS frameworks next.

Health Industry Cybersecurity Practices (HICP)

In collaboration with the healthcare sector, including the Department of **Health and Human Services** (**HHS**), the DHS supports the HICP. This framework offers cybersecurity guidelines and best practices specifically designed for healthcare organizations to safeguard patient data and medical systems.

Electricity Subsector Cybersecurity Capability Maturity Model (ES-C2M2)

The DHS, in collaboration with the electricity subsector, developed the ES-C2M2 framework. It assesses and improves the cybersecurity capabilities of the energy sector, focusing on electric utilities and grid operators. This framework helps organizations in the energy sector enhance their cybersecurity defenses.

Pipeline security guidelines

While not a formal framework, the DHS provides guidelines and resources to enhance the security of pipeline infrastructure. These guidelines help protect the transportation and distribution of critical resources, such as oil and natural gas.

Chemical Facility Anti-Terrorism Standards (CFATS)

CFATS is primarily focused on physical security but includes cybersecurity considerations. It mandates that chemical facilities integrate cybersecurity into their overall security plans to mitigate cyber threats impacting physical safety.

Information Sharing and Analysis Centers (ISACs)

DHS supports and collaborates with various ISACs, which are sector-specific organizations that facilitate information sharing and threat intelligence sharing among organizations within critical infrastructure sectors. ISACs often provide sector-specific guidance and resources related to cybersecurity.

HITRUST CSF

The **Health Information Trust (HITRUST) Common Security Framework (CSF)** is a comprehensive framework for managing information security and privacy risks in healthcare organizations and their business associates. HITRUST CSF was originally developed to address the complex regulatory and security challenges associated with the healthcare industry, which handles sensitive **patient health information (PHI)** and **electronic health records (EHRs)**. The framework was developed by the **HITRUST Alliance**, a private-sector organization.

The primary purpose of the HITRUST CSF is to provide healthcare organizations with a standardized and prescriptive framework for managing cybersecurity and privacy risks effectively. It helps organizations protect sensitive health information and comply with various regulations, including the **Health Insurance Portability and Accountability Act (HIPAA)**.

CIS Controls

The **Center for Internet Security (CIS) Controls**, formerly known as the **SANS Top 20 Critical Security Controls**, is a set of best practices and security guidelines designed to help organizations strengthen their cybersecurity posture. These controls cover areas such as asset management, continuous monitoring, and data protection and provide a prioritized and actionable framework for reducing cybersecurity risks. They are particularly relevant to critical infrastructure sectors, where the consequences of cyberattacks can have far-reaching impacts on public safety, national security, and the economy.

The CIS Controls are organized into three implementation groups, each with a different focus, as shown in the following table:

CIS Control group	Description
Basic Cyber Hygiene (CIS Controls 1-6)	This group focuses on fundamental security measures that organizations should implement to establish a strong cybersecurity foundation. It includes controls such as inventory and control of hardware assets, inventory and control of software assets, continuous vulnerability assessment and remediation, and controlled use of administrative privileges.

CIS Control group	Description
Foundational Security Controls (CIS Controls 7-16)	These controls build upon the basics and address more advanced security measures. They cover areas such as email and web browser protections, data protection, secure configuration, and security training and awareness. These controls are essential for organizations looking to establish a robust security program.
Organizational-Level Controls (CIS Controls 17-20)	This group focuses on organizational-level security practices, including incident response and management, continuous monitoring, and security metrics. These controls help organizations proactively detect, respond to, and recover from security incidents.

Table 6.11 – CIS Controls

It's important to note that the choice of security policies and frameworks should be based on the specific sector and regulatory requirements governing the critical infrastructure in question. Additionally, a comprehensive security strategy should involve ongoing risk assessments, regular audits, and a commitment to continuous improvement to adapt to evolving cyber threats.

Summary

In this chapter, we focused on network security and continuous monitoring techniques and gained valuable insights into the essential measures and policies required to safeguard critical infrastructure in an increasingly interconnected world.

Throughout this chapter, we delved into the intricate world of network security, understanding the fundamental principles that underpin the protection of critical systems. We explored the importance of robust authentication and access control mechanisms, the significance of encryption in securing data in transit, and the role of firewalls and IDSs in preventing unauthorized access.

Continuous monitoring emerged as a key theme, emphasizing the need for vigilant and proactive surveillance of network activities. We learned about the significance of real-time threat detection, and the utilization of SIEM systems to maintain a constant watchful eye on network behavior.

Policies and frameworks took center stage as we examined the comprehensive strategies and guidelines designed to fortify critical infrastructure. We explored the NIST Cybersecurity Framework and other industry-specific frameworks, understanding how they provide a structured approach to risk assessment, mitigation, and response.

In conclusion, this chapter served as a comprehensive guide to the world of network security and continuous monitoring techniques and it highlighted the critical role of policies and frameworks in protecting our critical infrastructure.

In the upcoming chapter, we will continue our exploration of safeguarding critical infrastructure by delving deeper into the areas of system security, endpoint protection, and application security.

References

To learn more about the topics that were covered in this chapter, take a look at the following resources:

- *Simpson, W. R., & Foltz, K. E. (2021). Network segmentation and zero trust architectures. In Lecture Notes in Engineering and Computer Science, Proceedings of the World Congress on Engineering (WCE) (pp. 201-206).*

- *Wagner, N., Şahin, C. Ş., Winterrose, M., Riordan, J., Pena, J., Hanson, D., & Streilein, W. W. (2016, December). Towards automated cyber decision support: A case study on network segmentation for security. In 2016 IEEE Symposium Series on Computational Intelligence (SSCI) (pp. 1-10). IEEE.*

- *Leischner, G., & Tews, C. (2007, April). Security through VLAN segmentation: Isolating and securing critical assets without loss of usability. In Proceedings of the 9th Annual Western Power Delivery and Automation Conference, Spokane, WA.*

- *Draney, B., Campbell, S., & Walter, H. (2007). NERSC Cyber Security Challenges That Require DOE Development and Support (No. LBNL-62284). Lawrence Berkeley National Lab.(LBNL), Berkeley, CA (United States).*

- *Bruzgiene, R., & Jurgilas, K. (2021). Securing remote access to information systems of critical infrastructure using two-factor authentication. Electronics, 10(15), 1819.*

- *Aloul, F., Al-Ali, A. R., Al-Dalky, R., Al-Mardini, M., & El-Hajj, W. (2012). Smart grid security: Threats, vulnerabilities and solutions. International Journal of Smart Grid and Clean Energy, 1(1), 1-6.*

- *Farquharson, J., Wang, A., & Howard, J. (2012, January). Smart grid cyber security and substation network security. In 2012 IEEE PES Innovative Smart Grid Technologies (ISGT) (pp. 1-5). IEEE.*

- *Anwar, A., & Mahmood, A. N. (2014). Cyber security of smart grid infrastructure. arXiv preprint arXiv:1401.3936.*

- *Mallouhi, M., Al-Nashif, Y., Cox, D., Chadaga, T., & Hariri, S. (2011, January). A testbed for analyzing security of SCADA control systems (TASSCS). In ISGT 2011 (pp. 1-7). IEEE.*

- *Peisert, S., Roberts, C., & Scaglione, A. (2020). Supporting Cyber Security of Power Distribution Systems by Detecting Differences Between.*

- *Bessani, A. N., Sousa, P., Correia, M., Neves, N. F., & Verissimo, P. (2008). The CRUTIAL way of critical infrastructure protection. IEEE Security & Privacy, 6(6), 44-51.*

- *Linda, O., Vollmer, T., & Manic, M. (2009, June). Neural network based intrusion detection system for critical infrastructures. In 2009 international joint conference on neural networks (pp. 1827-1834). IEEE.*

- *Ashoor, A. S., & Gore, S. (2011). Importance of intrusion detection system (IDS). International Journal of Scientific and Engineering Research, 2(1), 1-4.*

- *Premaratne, U. K., Samarabandu, J., Sidhu, T. S., Beresh, R., & Tan, J. C. (2010). An intrusion detection system for IEC61850 automated substations. IEEE Transactions on Power Delivery, 25(4), 2376-2383.*

- *Radoglou-Grammatikis, P. I., & Sarigiannidis, P. G. (2019). Securing the smart grid: A comprehensive compilation of intrusion detection and prevention systems. Ieee Access, 7, 46595-46620.*

- *Mohan, S. N., Ravikumar, G., & Govindarasu, M. (2020, October). Distributed intrusion detection system using semantic-based rules for SCADA in smart grid. In 2020 IEEE/PES Transmission and Distribution Conference and Exposition (T&D) (pp. 1-5). IEEE.*

- *Adepu, S., & Mathur, A. (2018). Distributed attack detection in a water treatment plant: Method and case study. IEEE Transactions on Dependable and Secure Computing, 18(1), 86-99.*

- *Raciti, M., Cucurull, J., & Nadjm-Tehrani, S. (2012). Anomaly detection in water management systems. In Critical Infrastructure Protection: Information Infrastructure Models, Analysis, and Defense (pp. 98-119). Berlin, Heidelberg: Springer Berlin Heidelberg.*

- *Ramotsoela, T. D., Hancke, G. P., & Abu-Mahfouz, A. M. (2020). Behavioural intrusion detection in water distribution systems using neural networks. IEEE Access, 8, 190403-190416.*

- *Gao, W., Morris, T., Reaves, B., & Richey, D. (2010, October). On SCADA control system command and response injection and intrusion detection. In 2010 eCrime Researchers Summit (pp. 1-9). IEEE.*

- *Iqbal, M., & Riadi, I. (2019). Analysis of Security Virtual Private Network (VPN) using OpenVPN. International Journal of Cyber-Security and Digital Forensics, 8(1), 58-65.*

- *Tyson, J. (2001). How Virtual Private Networks Work. Howstuffworks (Jul. 12, 2005).*

- *Abdelghani, T. (2019). Implementation of defense in depth strategy to secure industrial control system in critical infrastructures. American Journal of Artificial Intelligence, 3(2), 17-22.*

- *Shah, S., & Mehtre, B. M. (2015). An overview of vulnerability assessment and penetration testing techniques. Journal of Computer Virology and Hacking Techniques, 11, 27-49.*

- *Denis, M., Zena, C., & Hayajneh, T. (2016, April). Penetration testing: Concepts, attack methods, and defense strategies. In 2016 IEEE Long Island Systems, Applications and Technology Conference (LISAT) (pp. 1-6). IEEE.*

- *Hoffmann, J. (2015, April). Simulated penetration testing: From" dijkstra" to" turing test++". In Proceedings of the international conference on automated planning and scheduling (Vol. 25, pp. 364-372).*

- *Lyon, G. F. (2009). Nmap network scanning: The official Nmap project guide to network discovery and security scanning. Insecure.*

- *Coffey, K., Smith, R., Maglaras, L., & Janicke, H. (2018). Vulnerability analysis of network scanning on SCADA systems. Security and Communication Networks, 2018.*

- *Allman, M., Paxson, V., & Terrell, J. (2007, October). A brief history of scanning. In Proceedings of the 7th ACM SIGCOMM conference on Internet measurement (pp. 77-82).*

- *Zhang, L., & Thing, V. L. (2021). Three decades of deception techniques in active cyber defense-retrospect and outlook. Computers & Security, 106, 102288.*

- *Stoll, Cliff. The cuckoo's egg: tracking a spy through the maze of computer espionage. Simon and Schuster, 2005.*

- *Stoll, C. (1988). Stalking the wily hacker. Communications of the ACM, 31(5), 484-497.*

- *Stafford, V. A. (2020). Zero trust architecture. NIST special publication, 800, 207.*

- *D'Silva, D., & Ambawade, D. D. (2021, April). Building a Zero Trust Architecture Using Kubernetes. In 2021 6th International Conference for Convergence in Technology (i2ct) (pp. 1-8). IEEE.*

- *Federici, F., Martintoni, D., & Senni, V. (2023). A zero-trust architecture for remote access in industrial IoT infrastructures. Electronics, 12(3), 566.*

- *Vallentin, M., Sommer, R., Lee, J., Leres, C., Paxson, V., & Tierney, B. (2007). The NIDS cluster: Scalable, stateful network intrusion detection on commodity hardware. In Recent Advances in Intrusion Detection: 10th International Symposium, RAID 2007, Gold Coast, Australia, September 5-7, 2007. Proceedings 10 (pp. 107-126). Springer Berlin Heidelberg.*

- *Jacobson, V., Leres, C., & McCanne, S. (1989). The tcpdump manual page. Lawrence Berkeley Laboratory, Berkeley, CA, 143, 117.*

- *McCanne, S., & Jacobson, V. (1993, January). The BSD Packet Filter: A New Architecture for User-level Packet Capture. In USENIX winter (Vol. 46, pp. 259-270).*

- *Paxson, V. (1999). Bro: a system for detecting network intruders in real-time. Computer networks, 31(23-24), 2435-2463.*

- *Stoffer, V., Sharma, A., & Krous, J. (2015). 100g intrusion detection. Lawrence Berkeley National Laboratory (LBL).*

- *Chaudet, C., Fleury, E., Lassous, I. G., Rivano, H., & Voge, M. E. (2005, October). Optimal positioning of active and passive monitoring devices. In Proceedings of the 2005 ACM conference on Emerging network experiment and technology (pp. 71-82).*

- *Shaneman, K., & Gray, S. (2004, October). Optical network security: technical analysis of fiber tapping mechanisms and methods for detection & prevention. In IEEE MILCOM 2004. Military Communications Conference, 2004. (Vol. 2, pp. 711-716). IEEE.*

- *Svoboda, J., Ghafir, I., & Prenosil, V. (2015). Network monitoring approaches: An overview. Int J Adv Comput Netw Secur, 5(2), 88-93.*

- *Agarwal, D., González, J. M., Jin, G., & Tierney, B. (2003). An infrastructure for passive network monitoring of application data streams.*

- Yan, J., Liu, C. C., & Govindarasu, M. (2011, March). *Cyber intrusion of wind farm SCADA system and its impact analysis*. In 2011 IEEE/PES power systems conference and exposition (pp. 1-6). IEEE.

- Jhatakia, D. (2014). *Network Packet Monitoring Optimizations in Data Centre*. Happiest Minds Technologies, March.

- Kim, H., Chen, X., Brassil, J., & Rexford, J. (2021). *Experience-driven research on programmable networks*. ACM SIGCOMM Computer Communication Review, 51(1), 10-17.

- Liang, J., & Kim, Y. (2022, January). *Evolution of firewalls: Toward securer network using next generation firewall*. In 2022 IEEE 12th Annual Computing and Communication Workshop and Conference (CCWC) (pp. 0752-0759). IEEE.

- Jaggernauth, E., & Rocke, S. (2021). *Effectiveness of Paired Next Generation Firewalls in Securing Industrial Automation and Control Systems: A Case Study*. West Indian Journal of Engineering, 44(1).

- Sadeghian, A., & Zamani, M. (2014, February). *Detecting and preventing DDoS attacks in botnets by the help of self triggered black holes*. In 2014 Asia-Pacific Conference on Computer Aided System Engineering (APCASE) (pp. 38-42). IEEE.

- Disterer, G. (2013). *ISO/IEC 27000, 27001 and 27002 for information security management*. Journal of Information Security, 4(2).

- International Organization for Standardization. (2022). *ISO/IEC 27001: Information technology — Security techniques — Information security management systems — Requirements (3rd ed.)*: https://www.iso.org/obp/ui/#iso:std:iso-iec:27001:ed-3:v1:en

- International Organization for Standardization. (2022). *ISO/IEC 27002: Information technology — Security techniques — Code of practice for information security controls (3rd ed.)*: https://www.iso.org/obp/ui/#iso:std:iso-iec:27002:ed-3:v2:en

- Industrial Defender. (2023). *What is NERC CIP?*: https://www.industrialdefender.com/blog/what-is-nerc-cip

- Department of Homeland Security. (2022). *Critical infrastructure security and resilience R&D spend plan*: https://www.dhs.gov/sites/default/files/2022-06/ST%20-%20Critical%20Infrastructure%20Security%20and%20Resilience%20RDTE%20Spend%20Plan.pdf

- Kusnezov,D., (2023,). *Protecting our critical infrastructure during uncertain times*. Department of Homeland Security: https://www.dhs.gov/science-and-technology/news/2023/11/01/protecting-our-critical-infrastructure-during-uncertain-times

- *Cybersecurity & Infrastructure Security Agency. (2020). Transportation Systems Sector Cybersecurity Framework Implementation Guide*: `https://www.cisa.gov/resources-tools/resources/transportation-systems-sector-cybersecurity-framework-implementation-guide`

- *Healthcare and Public Health Sector Coordinating Council Cybersecurity Working Group. (2023). Health Industry Cybersecurity Practices: Managing Threats and Protecting Patients*: `https://405d.hhs.gov/Documents/HICP-Main-508.pdf`

- *U.S. Department of Energy. (2022). Cybersecurity Capability Maturity Model (C2M2) version 2.1. Retrieved [date you accessed the document]*: `https://www.energy.gov/sites/default/files/2022-06/C2M2%20Version%202.1%20June%202022.pdf`

- *Transportation Security Administration. (2018). Pipeline security guidelines*: `https://www.tsa.gov/sites/default/files/pipeline_security_guidelines.pdf`

- *National Council of ISACs. (n.d.). Member ISACs*: `https://www.nationalisacs.org/member-isacs-3`

- *HITRUST Framework*: `https://hitrustalliance.net/hitrust-framework`

- *Center for Internet Security. (n.d.). CIS Controls list*: `https://www.cisecurity.org/controls/cis-controls-list`

7

Protecting Critical Infrastructure – Part 2

This chapter follows the foundational groundwork established in the previous chapter, where we explored network security, continuous monitoring, and the development of security policies and frameworks. This chapter takes a deeper dive, building on how to protect critical infrastructure comprehensively, covering systems security and endpoint protection. It aims to provide an extensive understanding of how to safeguard the various components of critical infrastructure against sophisticated and evolving cyber threats. Emphasizing the importance of robust endpoint security, the effectiveness of antivirus and antimalware solutions, **endpoint detection and response** (**EDR**), and several aspects of application security, this chapter connects these elements to the broader context of digital security, offering insights into their integration within a holistic cybersecurity strategy. This continuation from the previous chapter underscores the complexity and interconnectivity of modern cyber defenses, highlighting the necessity of a multi-layered and nuanced approach to protecting vital digital assets.

In this chapter, we will cover the following topics:

- Systems security and endpoint protection
- Application security

Systems security and endpoint protection

Preventing breaches in systems is a crucial aspect of maintaining the integrity, confidentiality, and availability of data and systems.

Endpoint security, also known as endpoint protection or endpoint security management, refers to the practice of securing end-user devices such as computers, laptops, mobile devices, and servers from various security threats. The goal of endpoint security is to protect these devices from unauthorized access, data breaches, malware, and other cyber threats. Endpoints are often the entry points for cyberattacks, making it crucial to implement robust security measures at this level.

Let's examine a range of important topics that are essential for understanding and implementing endpoint and systems security solutions.

Antivirus/antimalware protection

Endpoint security solutions include antivirus and antimalware software to detect and remove malicious software from devices.

In the ever-evolving landscape of cybersecurity, *antivirus and antimalware protection* stand as stalwart guardians, defending digital ecosystems against malicious software threats. These technologies are the first line of defense, crucial in preserving the integrity and functionality of computing devices.

Antivirus programs are specifically engineered to identify, block, and eliminate software known as malware, which poses a threat to computer systems. Malware encompasses a broad spectrum of threats, including viruses, worms, trojan horses, spyware, ransomware, and more. The primary goal is to thwart unauthorized access, data theft, and system damage.

Antivirus solutions employ real-time scanning mechanisms, constantly monitoring files, applications, and networking activity. As soon as a potential threat is identified, the antivirus program takes swift action to neutralize or quarantine the malicious entity.

One of the traditional methods involves signature-based detection. Antivirus software maintains a database of known malware signatures, enabling it to recognize and eliminate familiar threats. Regular updates are critical to maintain the relevancy of the database against new and evolving threats.

In addition to signature-based detection, modern antivirus solutions leverage heuristic analysis. This involves examining the behavior of files and programs and identifying patterns or characteristics indicative of malicious intent. This proactive approach helps combat new and previously unseen threats.

Advanced antivirus tools incorporate behavioral monitoring, observing the behavior of applications in real time. Any deviation from normal behavior triggers alerts or actions to prevent potential threats from causing harm.

Antivirus protection has evolved to leverage cloud-based **threat intelligence** (**TI**). By tapping into a vast network of information, antivirus programs can access real-time data on emerging threats, enhancing their ability to detect and neutralize previously unknown malware.

Automated updates ensure that the software is equipped to handle the latest threats, vulnerabilities, and attack vectors, reinforcing the security posture of the entire system.

Antivirus and antimalware protection serve as a foundational element in the broader cybersecurity strategy. While not a standalone solution, they complement other security measures, such as firewalls, **intrusion detection systems** (**IDS**), and user education, forming a robust defense against multifaceted cyber threats.

While antivirus and antimalware software are essential components of a robust cybersecurity strategy, there are specific scenarios where their use might not be recommended or may need careful consideration:

- **Incompatible systems**: Some older or specialized systems may not be compatible with modern antivirus software. In such cases, alternative security measures may need to be explored to ensure the continued protection of these systems.

- **Resource-intensive on low-end devices**: On devices with very limited resources (such as older computers or **Internet of Things (IoT)** devices), certain antivirus programs can be resource-intensive, leading to performance issues.

- **False positives impact critical systems**: In environments where false positives can have severe consequences, such as **industrial control systems (ICSs)**, the use of antivirus software may need to be carefully managed. False positives could potentially disrupt essential operations.

- **Compromised by advanced threats**: In certain high-security environments, particularly those dealing with **advanced persistent threats (APTs)**, relying solely on traditional antivirus solutions may be insufficient. Advanced threat actors may employ techniques that can bypass or evade conventional signature-based detection.

- **Custom or niche applications**: Some specialized or custom-built applications may have behaviors that trigger false positives or cause compatibility issues with antivirus software. In such cases, organizations may need to work with security experts to find a balanced solution.

- **Impact on system performance**: In environments where system performance is a critical factor, such as high-performance computing clusters, there might be concerns about the impact of antivirus scans on processing speeds. Organizations may need to balance security needs with system performance requirements.

- **Privacy concerns**: Some users, particularly in highly privacy-sensitive environments, might be concerned about the data collection practices of certain antivirus vendors. In such cases, selecting antivirus software with transparent privacy policies becomes crucial.

- **Security through obscurity**: In some niche or less-targeted environments, there might be a notion of *security through obscurity*, where the belief is that the systems are unlikely to be targeted. While relying on obscurity alone is not a recommended security practice, organizations might opt for alternative security measures based on their risk assessments.

- **Over-reliance on signature-based detection**: Antivirus solutions primarily relying on signature-based detection might be less effective against zero-day threats or polymorphic malware. In environments where these threats are a significant concern, a more comprehensive security strategy is advisable.

It's crucial to note that the decision to use or not use antivirus and antimalware software should be based on a thorough risk assessment and a clear understanding of the specific needs and constraints of the environment in question. In many cases, a combination of security measures, including regular software updates, user education, and network security controls, can enhance overall cybersecurity.

In a world where digital adversaries constantly innovate, antivirus and antimalware protection remain indispensable shields. Through their continual adaptation and vigilance, these technologies play a pivotal role in the ongoing battle for cybersecurity.

Firewalls

Firewalls are utilized to oversee and regulate the flow of data entering and leaving an endpoint device, playing a key role in thwarting unauthorized access and defending against attacks that are network-based.

Stateless firewalls operate by filtering network traffic based on static rules, examining each packet in isolation, and using criteria such as IP addresses and ports. This approach offers speed but limits security as it doesn't consider the context or state of the traffic. On the other hand, stateful firewalls enhance security by monitoring and remembering the state of active connections. They assess whether incoming packets are part of existing, new, or unsolicited connections, offering a more context-aware filtering process. While this results in greater security, it also introduces increased complexity and can potentially reduce performance compared to the more straightforward stateless firewalls.

An endpoint firewall, often referred to as a host-based firewall or personal firewall, is a security solution that operates at the individual device level, such as a computer, laptop, or mobile device. Unlike network firewalls, discussed in the previous chapter, which protect entire networks, endpoint firewalls focus on safeguarding the specific device on which they are installed.

Endpoint firewalls, functioning as either standalone software or integral components of security suites, are designed to oversee traffic moving in and out of a specific device on a network. They serve as gatekeepers between the device and the broader network, managing communications in accordance with established security protocols.

The primary purpose of an endpoint firewall is to protect the device from unauthorized access, malware, and other network-based threats. It does this by examining data packets entering or leaving the device and making decisions based on a set of predetermined rules. Endpoint firewalls filter network traffic based on various criteria, such as IP addresses, port numbers, and protocols. This filtering capability allows the firewall to permit or block specific types of traffic, helping to prevent malicious connections.

Many modern endpoint firewalls include application control features, allowing users to define rules for individual applications or processes. This helps in preventing unauthorized or potentially malicious applications from accessing the network. Endpoint firewalls actively monitor network connections established by applications on the device. If an application attempts to create an unauthorized connection or exhibits suspicious behavior, the firewall can intervene and block the connection.

Users or administrators can configure the firewall settings through rule-based policies. These rules dictate how the firewall should handle different types of network traffic, ensuring that the device's network communication aligns with security policies.

Here are some examples of firewall rules for application control on an endpoint firewall:

Rule ID	Rule Description	Conditions	Action
1	Allow inbound HTTP traffic	Incoming traffic on port 80	Allow
2	Block outbound unauthorized chat	Outbound traffic to unauthorized chat servers	Block
3	Allow inbound email services	Incoming traffic on ports 25, 110, 143	Allow
4	Block outbound peer-to-peer (P2P) file sharing	Outbound traffic to known P2P IP ranges	Block
5	Allow inbound VPN connections	Incoming traffic on VPN-specific ports	Allow
6	Block inbound traffic from malicious IPs	Incoming traffic from IPs on TI blacklist	Block
7	Allow outbound software updates	Outbound traffic to recognized update servers	Allow
8	Block inbound remote desktop access (RDP)	Incoming traffic on port 3389 (RDP)	Block
9	Allow outbound DNS queries	Outbound DNS requests on port 53	Allow
10	Block suspicious inbound traffic	Suspicious patterns detected in inbound traffic	Block

Table 7.1 – Firewall rule examples

Endpoint firewalls often employ stateful inspection, a technology that keeps track of the state of active connections. This allows the firewall to make context-aware decisions based on the current state of the connection, enhancing security. Some endpoint firewalls offer customizable security profiles that allow users to define different levels of security for various scenarios. For example, a user might configure a higher security profile when connected to a public Wi-Fi network.

Endpoint firewalls generate notifications or logs that provide information about blocked or permitted network traffic. These logs are valuable for monitoring security events and conducting post-incident analysis.

Incorporating endpoint firewalls into a multi-layered security approach adds a critical layer of defense at the device level. These firewalls are key in blocking unauthorized access, reducing malware impact, and managing application activities. Achieving a balance between robust security and maintaining user functionality is vital when setting up endpoint firewalls. Overly restrictive policies can hinder legitimate applications, while overly permissive settings may expose the device to security risks.

These types of firewalls are commonly recommended for user laptops and workstations because these devices are often more vulnerable to certain types of threats and have a higher likelihood of interacting with untrusted networks. Laptops and workstations are individual devices that are often used in various locations, both inside and outside an organization's network, such as coffee shops, airports, and public Wi-Fi networks. These networks may lack the same level of security controls as an organization's networks, making laptops more susceptible to various network-based threats. Endpoint firewalls help mitigate these risks by regulating network traffic on the device itself.

Also, with the increasing trend of remote work, user laptops often connect to home networks or public Wi-Fi, exposing them to different security challenges. Endpoint firewalls provide a crucial defense layer in these scenarios, helping to prevent unauthorized access and protect against potential threats. End users on laptops and workstations may inadvertently engage in risky behavior, such as downloading files from untrusted sources, clicking on malicious links, or using unauthorized applications. Endpoint firewalls can monitor and control these activities, providing an additional line of defense against user-related security incidents.

Workstations are susceptible to threats that can be introduced locally, such as removable media (USB drives) carrying malware. Endpoint firewalls can include device control features to manage the use of external storage devices and prevent the spread of threats via these vectors. User laptops and workstations are typically associated with specific individuals. Endpoint firewalls enable a more user-centric security approach by allowing policies and controls to be tailored to the specific needs and risks associated with each user.

Endpoint firewalls contribute to **data loss prevention** (DLP) efforts by controlling the flow of sensitive information from the device to the network. This is particularly crucial on user laptops and workstations that may handle sensitive data.

While endpoint firewalls are highly recommended for user laptops and workstations, it's important to note that they are just one component of a comprehensive security strategy. Network firewalls, IDSs and **intrusion prevention systems** (IPSs), secure web gateways, and other security measures at the network level complement the protection provided by endpoint firewalls. Together, these layers form a **defense-in-depth** (DiD) approach to cybersecurity.

In summary, an endpoint firewall is a critical component of endpoint security, providing a barrier against network-based threats and helping to ensure the security and integrity of individual devices within a network.

Host IDS/IPS

IDS and IPS are used to detect and respond to suspicious activities or potential security breaches on endpoints.

Host IDS/IPS are cybersecurity solutions designed to protect individual computing devices (hosts or endpoints) from security threats by monitoring and responding to suspicious activities. These systems play a crucial role in enhancing the security posture of individual hosts by detecting and, in the case of HIPS, actively preventing unauthorized or malicious activities. Let's look at their functions:

- **Host IDS**: The primary focus is on detecting security incidents and potential threats to individual hosts. It operates by analyzing system logs, file changes, network traffic, and other activities to identify patterns indicative of malicious behavior.

- **Host IPS**: In addition to detection, a HIPS includes preventive measures. It actively blocks or prevents known threats from affecting the system by taking actions such as blocking network connections, terminating malicious processes, or modifying system configurations.

Both HIDS and HIPS often leverage signature-based detection. This involves comparing observed patterns or behaviors on a host with known threat signatures. These signatures are typically based on previously identified malware characteristics or attack patterns.

Both systems utilize behavioral analysis to identify anomalies in the way applications and users interact with the system. Behavioral analysis helps in detecting deviations from normal patterns of behavior that may indicate a security threat.

HIDS and HIPS operate in real time, continuously monitoring events on the host. This real-time monitoring enables prompt detection of security incidents, allowing for timely responses to mitigate potential threats. Both systems analyze network traffic to and from the host, which includes monitoring for unusual patterns, identifying malicious connections, and detecting potential network-based attacks.

HIDS and HIPS often include file integrity monitoring capabilities. This involves tracking changes to critical system files and configurations. Unexpected modifications can be indicative of unauthorized access or the presence of malware.

Here's a table with examples of files and configurations that might be monitored:

Category	Examples of Monitored Files/Configurations
System binaries	`/bin/*`, `/sbin/*`, `/usr/bin/*` – Monitoring changes to essential binaries
Configuration files	`/etc/passwd`, `/etc/shadow` – Monitoring changes to user account information
Kernel modules	`/lib/modules/*` – Monitoring changes to kernel modules
System libraries	`/lib/*`, `/usr/lib/*` – Monitoring changes to critical system libraries

Category	Examples of Monitored Files/Configurations
Boot configuration	`/boot/grub/grub.cfg` – Monitoring changes to the bootloader configuration
Network configuration	`/etc/network/*` – Monitoring changes to network configuration files
System logs	`/var/log/*` – Monitoring changes to system log files
Critical applications	`/opt/application/*` – Monitoring changes to key application files
Security configuration	`/etc/security/*` – Monitoring changes to security-related configurations

Table 7.2 – Files and configuration examples

A host IDS generates alerts when it detects suspicious activities, providing security administrators with notifications to investigate potential security incidents. A host IPS, in addition to alerts, can take automated actions to block or prevent identified threats. Host IDS provides information that aids in **incident response** (**IR**) by generating alerts and supplying data for further investigation. Host IPS goes a step further by actively responding to incidents and automatically taking preventive actions to stop threats from causing harm.

Host IDS and Host IPS can function as standalone solutions dedicated to intrusion detection and prevention on individual hosts. They may also be integrated into broader security frameworks, complementing other security measures such as **network-based IDS** (**NIDS**) and firewalls.

Resource usage considerations vary based on the specific solution, but both HIDS and Host IPS aim to operate efficiently on individual hosts without causing significant performance degradation.

Host IDS and Host IPS, when deployed and configured appropriately, significantly contribute to an organization's DiD strategy by protecting individual hosts from a wide range of cybersecurity threats. Organizations often deploy these systems alongside other security measures to create a comprehensive security posture.

EDR

EDR solutions offer continuous monitoring, detection, and response functions on endpoint devices, enabling organizations to promptly identify and address security breaches.

EDR is a cybersecurity solution that focuses on detecting and mitigating advanced threats on individual computing devices. EDR goes beyond traditional antivirus and antimalware solutions by providing real-time visibility into endpoint activities and responding to security incidents promptly. Here's a table summarizing the key differences between Host IDS, Host IPS, and EDR:

Feature	Host IDS	Host IPS	EDR
Primary focus	Detection of security incidents	Prevention of security incidents	Detection, response, and investigation
Functionality	Detection with alerting	Prevention with immediate blocking	IR, forensic analysis, threat hunting
Signature-based detection	Yes	Yes	Yes
Behavioral analysis	Yes	Limited (primarily focused on signatures), although depends on the solution	Yes (strong emphasis on proactive analysis)
IR	Limited (alerting with investigation support)	Limited (may include basic response actions)	Comprehensive capabilities for response and remediation
Prevention emphasis	Detection and alerting	Prevention of known threats with blocking capabilities	Broad capabilities beyond prevention
Automated actions	Typically no automated prevention actions	Yes (automated blocking of known threats)	Comprehensive automation for IR
Integration with other security measures	Can be standalone or integrated	Often part of broader security frameworks	Often part of broader endpoint security solutions
Threat hunting	Limited (primarily focused on detection)	Limited (prevention-focused)	Supported for proactive identification of threats
Forensic analysis	Basic capabilities for incident investigation	Limited (may include basic forensics)	Robust tools for detailed forensic analysis
Real-time monitoring	Yes	Yes	Yes
Resource usage considerations	Resource considerations vary based on the solution implemented	Resource considerations vary based on the solution implemented	Resource considerations vary based on the solution implemented

Table 7.3 – HIDS, HIPS, and EDR key differences

Endpoint security is a critical component of a comprehensive cybersecurity strategy, especially as more organizations embrace remote work and mobile devices. As we have covered, it helps defend against a wide range of cyber threats and ensures that endpoints are secure entry points into the overall network.

Now, let's turn our attention to another critical aspect of our recommended security architecture: application security.

Application security

Application security refers to measures and practices designed to protect software applications from security threats, vulnerabilities, and unauthorized access. As applications play a central role in today's digital landscape, securing them is crucial to ensure the confidentiality, integrity, and availability of data, as well as to prevent unauthorized access and exploitation.

Let's explore together some key points on application security.

Secure software development life cycle

For cases where there is homebrew software being developed and deployed, implementing a **secure development lifecycle (SDL)** is key. Implementing a secure **software development life cycle (SDLC)** involves integrating security practices at every phase of the software development process. The following table shows the different phases that are involved in the SDLC and their key activities:

SDLC Phase	Key Activities
Requirements gathering	Identify security requirements
	Define authentication and authorization needs
	Risk assessment
Design phase	Conduct threat modeling
	Ensure secure architecture design
Implementation phase	Enforce secure coding standards
	Perform **static application security testing (SAST)**
	Use static code analysis tools
Testing phase	Perform **dynamic application security testing (DAST)**
	Conduct penetration testing
	Implement automated security testing

SDLC Phase	Key Activities
Deployment phase	Ensure secure deployment configurations
	Address environment security considerations
Post-deployment phase	Implement monitoring and IR
	Manage security patches effectively
Training and awareness	Provide developer training on secure coding practices
	Conduct security awareness programs
Compliance and documentation	Check compliance with security standards and regulations
	Maintain comprehensive documentation

Table 7.4 – SDLC key phases and activities

After examining the key points of the secure SDLC, let's shift our focus to understanding how conducting code reviews and static analysis can uncover security vulnerabilities.

Code reviews and static analysis

Regular code reviews and static code analysis tools help identify and address security vulnerabilities in the source code before the application is deployed.

Code reviews and static analysis are integral components of an SDL aimed at identifying and mitigating security vulnerabilities in the source code. Both practices are critical for ensuring that the code base is secure, resilient against potential threats, and aligns with established security standards. First, let's elaborate on code reviews in the context of application security:

Code Reviews		
Purpose	Identification of security flaws	Code reviews involve a manual examination of the source code by one or more developers. The primary purpose is to identify security flaws, programming errors, and adherence to coding standards.
	Knowledge transfer	Code reviews facilitate knowledge sharing among team members, promoting a collective understanding of the code base, its security requirements, and best practices.

Code Reviews		
Key practices	**Regular reviews**	Code reviews should be conducted regularly throughout the development process, with a focus on security-critical sections of the code.
	Checklist usage	Developers often use security checklists to ensure that common security issues, such as input validation, output encoding, and authentication, are appropriately addressed.
Security aspects in code reviews	**Input validation**	Ensure that user input is properly validated to prevent injection attacks such as SQL injection or **cross-site scripting (XSS)**.
	Authentication and authorization	Verify that authentication mechanisms are implemented securely and proper authorization controls are in place.
	Sensitive data handling	Review how sensitive data is handled, stored, and transmitted, ensuring encryption and secure storage practices.
	Error handling	Evaluate error-handling mechanisms to prevent information disclosure and enhance the resilience of the application.
Feedback and collaboration	**Constructive feedback**	Code reviews provide an opportunity for team members to offer constructive feedback on security-related issues, code quality, and potential improvements.
	Collaboration	Developers can collaborate during code reviews to share insights, discuss security considerations, and collectively make decisions to enhance the security of the code base.

Table 7.5 – Code reviews

Now, let's elaborate on static analysis in the context of application security:

Static Analysis		
Purpose	**Automated code analysis**	Static analysis involves the use of automated tools to analyze the source code without executing it. The primary purpose is to identify potential security vulnerabilities, coding errors, and adherence to coding standards.
	Early detection	Static analysis allows for the early detection of security issues during the development phase, reducing the likelihood of vulnerabilities making their way into the production environment.
Key practices	**Integration into CI/CD pipelines**	Static analysis tools can be integrated into **continuous integration/continuous deployment (CI/CD)** pipelines, enabling automated and consistent code analysis at each code commit.
	Custom rulesets	Organizations can customize rulesets based on their specific security policies, industry standards, and coding conventions.
Security aspects in static analysis	**Code vulnerability identification**	Static analysis tools identify potential vulnerabilities such as buffer overflows, injection flaws, and insecure cryptographic practices.
	Code complexity analysis	Assess code complexity, as complex code can lead to security vulnerabilities and make it harder to identify potential issues.
	Third-party component analysis	Analyze third-party libraries and components for known vulnerabilities and adherence to security best practices.
Automation and scalability	**Scalable analysis**	Static analysis tools can handle large code bases efficiently, making them suitable for projects of varying sizes.
	Continuous monitoring	The automated nature of static analysis allows for continuous monitoring of code changes, ensuring that new vulnerabilities are promptly identified.

Static Analysis		
Integration with development tools	**Integrated development environment (IDE) integration**	Some static analysis tools integrate directly into IDEs, providing developers with real-time feedback during the coding process.
	Issue tracking integration	Identified issues can be integrated into issue-tracking systems, streamlining the remediation process.
False positive management	**Adjusting rule thresholds**	Organizations may need to adjust rule thresholds to manage false positives and ensure that analysis results are actionable.
	Regular review	Teams should regularly review and validate static analysis findings to distinguish between true security issues and false positives.

Table 7.6 – Static analysis

Code reviews and static analysis complement each other in the pursuit of secure software development. Code reviews leverage human expertise to identify nuanced security issues and foster collaboration, while static analysis automates the identification of common vulnerabilities and ensures consistent analysis across the entire code base. Together, these practices contribute to building a robust security posture within the SDLC.

Authentication and authorization hardening

Establishing robust authentication processes for users, particularly through **multi-factor authentication (MFA)**, is critical to ensuring that only authorized individuals gain access to an application.

User authentication is the process of verifying the identity of an individual attempting to access a system or application. It involves presenting credentials, typically in the form of a username and password, to prove the user's identity. Users are required to provide unique credentials, usually a combination of a username and a password. These credentials serve as the initial proof of identity.

MFA enhances security by requiring users to present various forms of verification. This might include a known element (such as a password), a possessed item (such as a mobile device), or an inherent characteristic (such as a fingerprint).

Here are some ways to implement strong authentication:

- **Password policies**: Enforce strong password policies, including requirements for complex passwords.

- **Biometric authentication**: Where applicable and secure, incorporate biometric authentication methods such as fingerprint scanning, facial recognition, or authenticating users based on the unique characteristics of their voice.

- **Time-based one-time passwords (TOTPs)**: Implement a TOTP, a common form of MFA, which generates time-sensitive codes that users must input along with their passwords.

- **Device-based authentication**: Utilize device-based authentication, where access is granted only from recognized and authorized devices.

- **Smart cards and tokens**: These are physical devices that users possess and can use as an additional factor for authentication. Some examples are smart cards with embedded chips, USB security tokens, and hardware security keys.

- **Certificate-based authentication**: This form of authentication involves the use of digital certificates to verify the identity of users and devices; for example, **public key infrastructure (PKI)** certificates, X.509 certificates, and client certificates.

- **Single sign-on (SSO)**: SSO allows users to access multiple applications with a single set of credentials, and strong authentication can be integrated to enhance security, such as combining SSO with MFA, biometric authentication, or other strong authentication methods.

- **Authentication based on location**: Authentication based on the geographical location of the user, adding an additional layer of context; for example, geofencing, IP address verification, and GPS-based authentication.

- **Personal identification number (PIN)**: Using a combination of a PIN and a password for stronger authentication.

Organizations often choose a combination of these authentication methods based on their security requirements, the nature of the application, and user experience considerations. The key is to strike a balance between security and usability while considering the specific needs of the users and the organization's risk profile.

Implementing proper authorization controls ensures that users have appropriate permissions to access specific resources or perform certain actions within the application.

Authorization is the process of determining what actions or operations a user, system, or application can perform within a system or resource. It defines and enforces access policies, specifying the level of access granted to authenticated entities based on their roles, permissions, or attributes. Strong authorization controls are crucial for maintaining the confidentiality, integrity, and availability of sensitive information and resources. The following table shows a few examples of implementing strong authorization controls:

Authorization Control	Definition	Example
Role-based access control (RBAC)	Assigns roles to users and grants permissions based on those roles.	In a healthcare system, a nurse's role may have permission to view patient records, while a doctor's role may have permission to update patient information.
Attribute-based access control (ABAC)	Uses attributes (user characteristics, resource properties) to make access control decisions.	An e-commerce platform grants access to customer order history based on the `customer_type` attribute (for example, premium customers versus regular customers).
Mandatory access control (MAC)	Sets access controls based on labels or classifications assigned to subjects and objects.	Military or government systems use MAC to restrict access based on security clearances.
Discretionary access control (DAC)	Allows users to control access to their own resources.	Filesystems often use DAC, where the owner of a file can specify who has read, write, or execute permissions on that file.
Rule-based access control	Uses rules or policies to determine access permissions.	A financial application may have rules preventing a user from making transactions above a certain amount without additional approvals.
Time-based access control	Restricts access to certain time periods.	A company's network may only allow software updates during specific maintenance windows.
Attribute-based encryption (ABE)	Encrypts data based on attributes, and access is granted to those with matching attributes.	A document encrypted with ABE may only be accessible to users with specific roles or attributes.
Policy-based access control	Defines access rules based on policies set by administrators.	An organization may have a policy that restricts access to sensitive data to employees with a specific job role and completion of a privacy training course.

Authorization Control	Definition	Example
Fine-grained access control	Provides precise control over access permissions at a granular level.	An ICS may grant different levels of access to operators based on their expertise and responsibilities.
Audit trails and monitoring	Tracks and logs access events for later review.	A financial institution may monitor user activities and generate audit reports to ensure compliance with regulatory requirements.
Least privilege principle	Grants only the minimum level of access required to perform tasks.	A database administrator may have full access to a database, while a customer support representative has only read access to specific tables.

Table 7.7 – Authorization controls

It's essential to underscore the importance of implementing a multi-layered security approach. Integrating diverse authentication methods and robust authorization controls not only fortifies an organization's defense against unauthorized access but also enhances the overall security posture. By prioritizing both the security and usability aspects, organizations can ensure that their systems remain resilient, secure, and accessible in an increasingly complex digital landscape.

Data encryption

Encrypting data is a crucial security practice that turns clear, readable information into a secure, encoded format, effectively blocking unauthorized access. This method is essential in protecting confidential information, not only during its transfer but also while it is stored, ensuring its safety both in transit and at rest. Let's elaborate on the two aspects mentioned.

Data-in-transit encryption

Data in transit refers to information that is actively moving from one location to another, typically over a network. Data-in-transit encryption ensures that the data remains confidential and integral during its journey between a user's device and the application server.

Let's examine some data-in-transit encryption key points:

Key Points	Details
Secure transmission protocols	Encrypts data in transit using protocols such as HTTPS, which employs **Transport Layer Security (TLS)** to secure communication between clients and servers
Confidentiality	Ensures confidentiality by preventing interception of sensitive information during transmission
Integrity	Maintains data integrity through cryptographic techniques, detecting and preventing unauthorized modifications

Table 7.8 – Data-in-transit encryption key points

Now, let's learn ways of implementing data-in-transit encryption:

Implementation	Description	Use Case
HTTP Secure (HTTPS)	Extension of HTTP using TLS for secure web communication	Securing communication between web browsers and websites
TLS	A cryptographic protocol that ensures privacy and integrity in communication	Email protocols (**Simple Mail Transfer Protocol (SMTP)**, **Internet Message Access Protocol (IMAP)**, **Post Office Protocol 3 (POP3)**, VPNs, and various applications requiring secure data transfer
Internet Protocol Security (IPsec)	A suite of protocols for authenticating and encrypting data at the IP layer	Securing communication between network devices in VPNs
Secure Shell (SSH)	Cryptographic protocol for secure remote access and administration of network devices and hosts	Securely managing servers and network devices over an unsecured network
Secure/Multipurpose Internet Mail Extensions (S/MIME)	Standard for secure email communication using encryption and digital signatures	Ensuring the confidentiality and authentication of email messages

Implementation	Description	Use Case
Pretty Good Privacy (PGP)	Program for data encryption and decryption, combining symmetric and public-key cryptography	Securing email communication and encrypting files
Datagram TLS (DTLS)	A variant of TLS for securing communication over datagram protocols such as **User Datagram Protocol (UDP)**	Secure communication in scenarios where connection-oriented protocols are not feasible, such as real-time applications
Wi-Fi Protected Access 3 (WPA3)	Wi-Fi security standard with modern cryptographic protocols for enhanced wireless security	Securing Wi-Fi networks to protect against unauthorized access and attacks

Table 7.9 – Data-in-transit encryption implementation

After learning about the various methods of encrypting data in transit, it becomes evident that implementing robust encryption techniques is crucial for ensuring the security and integrity of sensitive information as it moves across networks.

Data-at-rest encryption

Data at rest refers to information stored on a physical or digital medium, such as databases, files, or storage devices. Encryption of data at rest safeguards stored information by preventing unauthorized access, particularly in scenarios of physical theft or unauthorized entry into storage systems.

Let's examine some data-at-rest encryption key points:

Key Points	Details
Protection of stored data	Applies encryption algorithms to data before storage, safeguarding information on databases, filesystems, and storage devices
Securing files	Implements encryption for repositories storing sensitive information, complying with industry regulations and data protection laws
Compliance requirements	Meets legal and regulatory requirements, such as the **Health Insurance Portability and Accountability Act (HIPAA)** and the **General Data Protection Regulation (GDPR)**, by implementing data-at-rest encryption as part of security measures

Table 7.10 – Data-at-rest encryption key points

Let's now learn some ways of implementing data encryption at rest:

Implementation	Description	Use Case
Full disk encryption (FDE)	Encrypts the entire storage device, ensuring that all data on the disk is protected	Securing data on laptops, desktops, and servers to prevent unauthorized access in case of theft or unauthorized access
File-level encryption	Encrypts individual files or directories, providing granular control over encrypted data	Protecting specific sensitive files or folders, often used in collaboration platforms, cloud storage, or file-sharing systems
Database encryption	Encrypts data stored in databases, protecting sensitive information within the database	Safeguarding sensitive information in databases, including financial records, customer data, and other critical business data
Cloud storage encryption	Encrypts data stored in cloud storage services, ensuring security in the cloud environment	Ensuring the confidentiality of data stored in cloud platforms, preventing unauthorized access to files and documents
Tape encryption	Encrypts data stored on magnetic tapes, commonly used for backup and archival purposes	Protecting sensitive data stored on tapes, ensuring confidentiality during long-term storage and backup processes
Network-attached storage (NAS) encryption	Encrypts data on NAS devices to secure shared data in a network	Ensuring the security of shared data on NAS devices, protecting against unauthorized access from network users
Storage area network (SAN) encryption	Encrypts data on SANs, enhancing security in enterprise storage environments	Securing data within enterprise-level SANs, providing an additional layer of protection for critical business data
Virtual disk encryption	Encrypts virtual disks or volumes, providing security for virtualized environments	Protecting data in virtualized environments, securing virtual disks used in VMs and cloud computing platforms

Implementation	Description	Use Case
Hardware-based encryption	Uses specialized hardware components to perform encryption and decryption processes	Enhancing data security through dedicated hardware, often used in self-encrypting drives (SEDs) and hardware security modules (HSMs)
Application-layer encryption	Encrypts data within specific applications, offering encryption features at the application level	Providing data security within applications, commonly used in software that handles sensitive information such as password managers

Table 7.11 – Data-at-rest encryption implementation

These implementations of data encryption at rest are crucial for protecting sensitive information stored on various platforms and storage mediums. Each has its own strengths and use cases, and organizations often employ a combination of these methods based on their specific security requirements.

Session management

Session management is a critical component of application security that involves the creation, maintenance, and termination of user sessions. A session begins when a user logs in to an application and ends when the user logs out or after a defined period of inactivity. Proper session management is essential to prevent various security vulnerabilities and protect user data. Here are key aspects of session management for application security:

Session Management Practices	Description	Importance
Session creation	Secure authentication practices Unique and randomly generated **session identifiers (SIDs)**	Ensures secure user login and provides unique SIDs to prevent predictability or unauthorized access
Session maintenance	Appropriate session timeouts Dynamic extension based on user activity	Balances security and user convenience by defining session durations and dynamically adjusting based on user interaction
Secure session handling	Use of secure session tokens Regeneration of SIDs after login	Prevents session hijacking, fixation attacks, and unauthorized access through secure tokens and identifier regeneration

Session Management Practices	Description	Importance
TLS	Usage of HTTPS to encrypt the entire communication between client and server	Protects session data from eavesdropping and **person-in-the-middle** (PITM) attacks by encrypting communication
XSS mitigation	Application of secure cookie attributes Input validation to prevent script injection	Mitigates XSS attacks targeting session cookies and ensures that user input is sanitized to prevent malicious script injection
Session revocation	User-initiated logout Implementation of mechanisms for session revocation	Allows users to manually log out and provides mechanisms to promptly revoke sessions in case of suspicious activity or logout
Session monitoring and logging	Maintenance of audit trails Implementation of anomaly detection mechanisms	Provides a comprehensive record of user session activities, and detects and responds to unusual or suspicious session behavior
User session information	Storage of limited and essential information in session data Avoidance of sensitive data storage in session variables	Reduces the risk of data exposure by storing only necessary information in session data
Session testing	Regular security testing, including penetration testing and vulnerability assessments	Identifies and addresses potential session-related vulnerabilities through proactive security testing

Table 7.12 – Session management key aspects

In summary, session management is vital for application security, encompassing everything from secure session creation to thorough monitoring and testing. These practices are crucial for protecting user data against unauthorized access and various security threats, thereby maintaining the integrity and trust of the application. Regular updates and tests are essential to keep these defenses strong against evolving challenges.

Security patching and updates

Security patching and updates are vital components of a robust cybersecurity strategy. They involve the timely application of patches and updates to software, operating systems, frameworks, libraries, and other components to address known vulnerabilities and enhance overall security.

Let's explore security patching best practices:

Security Patching and Updates	Description	Importance
Timely updates	Involves applying security patches and updates promptly to software components, including the operating system, frameworks, libraries, and third-party modules.	**Vulnerability mitigation** Reduces the risk of exploitation by addressing known vulnerabilities
		Enhanced security posture Maintains a hardened and up-to-date software environment, contributing to overall security
		Compliance requirements Meets regulatory standards that mandate the timely application of security patches
		Protection against exploits Addresses vulnerabilities that, if exploited, could lead to unauthorized access and data breaches
		Minimized attack surface Closes off known entry points, reducing the overall attack surface
		Security incident prevention Proactive measures to prevent security incidents and data breaches
		Vendor support and end-of-life (EOL) concerns Ensures ongoing vendor support and addresses EOL concerns

Key Practices		
Regular vulnerability assessment	Conduct regular assessments to identify potential security vulnerabilities in the software stack.	**Proactive identification** Identifies potential vulnerabilities before they can be exploited
Monitoring security advisories	Stay informed about security advisories and updates released by software vendors, open source communities, and security organizations.	**Timely awareness** Keeps the organization aware of emerging threats and the availability of patches
Patch management system	Implement a system to automate the identification, testing, and application of patches across the organization.	**Efficient management** Automates the patching process, ensuring efficient identification and deployment
Prioritization of critical patches	Prioritize the application of critical patches that address known vulnerabilities with a high risk of exploitation.	**Risk mitigation** Focuses efforts on addressing the most critical vulnerabilities first
Scheduled patching cycles	Establish scheduled patching cycles to consistently apply updates without causing undue disruption to business operations.	**Consistent updates** Ensures a regular and consistent approach to applying security patches
Dependency scanning	Regularly scan for dependencies in applications, including libraries and third-party modules, and ensure they are updated to address vulnerabilities.	**Comprehensive security** Ensures that dependencies are also secure, reducing the overall risk of vulnerabilities

Challenges		
Compatibility issues	Updates may introduce compatibility issues with existing applications, requiring thorough testing before deployment.	**Thorough testing required** Ensures that updates do not disrupt existing applications or services
Downtime concerns	Applying updates may require system restarts or downtime, necessitating careful planning to minimize disruption.	**Strategic planning** Planning for downtime ensures that updates are applied with minimal impact on business operations
Resource intensiveness	Managing updates for a large and complex IT environment can be resource-intensive, requiring efficient patch management systems.	**Efficient resource allocation** Requires effective systems and processes to manage updates across a large IT infrastructure
Critical system dependencies	Some critical systems may have dependencies on specific software versions, making updates more challenging.	**Strategic approach** Requires a strategic approach to managing updates for critical systems with specific dependencies

Table 7.13 – Security patching best practices

In summary, security patching and updates are crucial for maintaining a strong cybersecurity defense. By promptly applying updates, regularly assessing vulnerabilities, and efficiently managing patches, organizations can mitigate risks and strengthen their security posture. Addressing challenges such as compatibility and downtime with strategic planning is essential. This proactive approach is key to protecting against the ever-evolving landscape of cyber threats.

Penetration testing

In the expansive realm of cybersecurity, the practice of penetration testing, often synonymous with ethical hacking or pen testing, emerges as a pivotal methodology. This approach involves skilled professionals simulating real-world cyberattacks, adopting the perspective of malicious actors seeking to identify vulnerabilities within applications, networks, or systems. The overarching goal is to comprehensively assess the security posture by deliberately attempting to exploit weaknesses in a controlled and ethical manner.

Within the domain of penetration testing, several key practices come into play. A critical initial step involves the clear definition of the test's scope, delineating the target systems, applications, and the permissible extent to which the testing team can venture in their pursuit to exploit vulnerabilities.

In white-box penetration testing, the tester is given complete access to the system's internal data, which includes code base, documentation, and network information, allowing for a thorough assessment from an insider's perspective. Black-box penetration testing, on the other hand, involves the tester working with no prior knowledge of the system, akin to an attacker from the outside, which can reveal vulnerabilities that may be exploited by unauthorized users. Lastly, gray-box penetration testing offers a middle ground where the tester has some limited knowledge, such as user-level access or system layouts, providing a more realistic scenario of how an informed insider might exploit system weaknesses.

Simulation of diverse cyberattacks takes center stage, mimicking the strategies a malicious actor might employ. This includes, but is not limited to, SQL injection, XSS, privilege escalation, and network-based attacks. The testing arsenal encompasses a spectrum of automated tools alongside manual testing techniques. While automated tools efficiently identify common vulnerabilities, manual testing allows for a more in-depth analysis.

Skilled professionals meticulously identify and document vulnerabilities susceptible to exploitation. This encompasses both technical vulnerabilities in the code and configuration errors within the system. Each identified vulnerability undergoes a rigorous risk assessment, considering factors such as potential impact, likelihood of exploitation, and ease of mitigation.

Various exploitation techniques are employed to vividly demonstrate the real-world impact of identified vulnerabilities. This aids stakeholders in comprehending the severity of the issues at hand. Upon completion, a detailed report is compiled, featuring a comprehensive list of vulnerabilities, their potential impact, and recommendations for remediation.

The importance of penetration testing cannot be overstated. It transcends theoretical assessments by simulating real-world attack scenarios, providing invaluable insights into vulnerabilities that may be exploited by malicious actors. By pinpointing and addressing vulnerabilities, penetration testing becomes a potent technique for mitigating the risk of security breaches, data leaks, and unauthorized access.

Many regulatory standards and compliance frameworks mandate regular penetration testing as an integral facet of security best practices. It serves as an educational tool, raising awareness among stakeholders about potential security risks and the importance of maintaining a robust security posture.

The iterative nature of penetration testing establishes a feedback loop for continuous improvement. Organizations gain a deeper understanding of their security weaknesses, enabling proactive measures to enhance security. Insights garnered from penetration testing enable organizations to prioritize remediation efforts based on the severity and impact of identified vulnerabilities.

Demonstrating a commitment to security through penetration testing builds trust with users, customers, and other stakeholders who rely on the integrity and security of the system. Yet, penetration testing is not without its challenges. Testing tools may generate false positives, necessitating manual verification to distinguish between actual vulnerabilities and false alarms.

The scope of penetration testing may be limited, potentially leaving certain vulnerabilities undetected if they fall outside the defined scope. Conducting thorough penetration tests can be resource-intensive, demanding both time and skilled personnel. In some instances, activities associated with penetration testing may impact the availability or performance of production systems, requiring careful planning and coordination.

Penetration testing stands as a proactive and essential security practice. It empowers organizations to identify and address vulnerabilities, ensuring the resilience of their systems against evolving cyber threats. As a critical component of a comprehensive security strategy, penetration testing remains indispensable in the ever-evolving landscape of cybersecurity.

Logging and monitoring

Within the domain of application security, the tandem of logging and monitoring emerges as a cornerstone for fortifying the resilience of software systems. These practices serve as vigilant custodians, capturing and analyzing events that unfold within an application, thereby facilitating proactive threat detection, IR, and adherence to compliance standards.

In the case of event logging, the emphasis lies on implementing a comprehensive logging mechanism that diligently records security-relevant events. This encompasses a spectrum of activities, ranging from user authentication to access to sensitive data and modifications of configuration settings. The intrinsic value of this detailed logging becomes apparent in post-incident analysis, enabling security teams to reconstruct the sequence of events leading up to and during a security incident. Furthermore, these logs serve as invaluable artifacts for forensic investigations, aiding in the identification of root causes. From a compliance standpoint, they form a basis for auditing activities, ensuring alignment with regulatory standards.

Parallelly, the integration with **security information and event management** (**SIEM**) solution adds a layer of sophistication to the security landscape. SIEM systems serve as central hubs for real-time monitoring, analysis, and correlation of security events across an organization's infrastructure. By linking an application's logs with a SIEM system, organizations gain real-time insights into security events, enabling swift responses to potential threats. SIEM's capability to correlate events from various sources unveils patterns and anomalies that might remain elusive when scrutinized in isolation. Additionally, the system facilitates alerting and notification mechanisms, prompting security teams to respond promptly to emerging security incidents.

The integration with SIEM has many advantages. It elevates threat detection capabilities, providing a comprehensive view of activities across the application and infrastructure. Furthermore, it streamlines compliance efforts by generating reports that substantiate the presence and efficacy of security controls. The real-time monitoring capabilities of SIEM not only fortify IR but also contribute to a centralized and efficient log management system.

In essence, the symbiosis of logging and monitoring is foundational to a resilient application security strategy. By offering visibility into an application's behavior, these practices empower security teams to detect and respond to threats effectively. Simultaneously, they serve as custodians of compliance, maintaining meticulous records of security-relevant events, and thereby fortifying the overall security posture of the application.

IR and data recovery

The protection of data integrity and availability is foundational. A critical strategy in fortifying this resilience is the meticulous implementation of *data backup and recovery* measures.

Data backup involves the routine creation and maintenance of copies of crucial information at defined intervals. This proactive approach acts as a robust defense against potential data loss, whether stemming from accidental deletion, hardware failures, or malicious activities such as cyberattacks. The backup process entails capturing a comprehensive snapshot of the entire dataset or specific components essential for sustaining business operations.

Equally important is the establishment of a robust *recovery plan*. This plan intricately outlines the steps and procedures to be executed in the event of data loss or compromise. It extends beyond the technical nuances of data restoration, encompassing coordination and communication strategies for stakeholders involved in the recovery process. A well-defined recovery plan serves to minimize downtime and significantly contributes to the overall resilience of the application.

In application security, these practices are of immense significance. They act as a formidable mitigation strategy against the specter of data loss, ensuring that valuable information can be efficiently restored following an incident. In the face of cyber threats such as ransomware attacks, having up-to-date backups becomes a potent defense, allowing organizations to recover their data without succumbing to extortion demands and maintaining control over their information assets. Furthermore, a robust recovery plan plays a pivotal role in ensuring **business continuity** (**BC**) by minimizing the impact of data incidents, allowing organizations to swiftly resume operations with minimal disruption.

It's noteworthy that the intricacies of IR, including detailed strategies for handling data incidents and recovery, will be thoroughly explored in the upcoming chapter in the *IR* section. This section will get into things such as coordinated actions, communication protocols, and technical processes, providing a comprehensive guide for effectively managing and recovering from security incidents.

Application security is an ongoing process that requires a combination of technical measures, secure coding practices, and a proactive approach to identifying and addressing emerging threats. Organizations that prioritize application security reduce the risk of data breaches, protect user privacy, and maintain the trust of their customers and stakeholders.

Summary

This chapter imparted key insights into various aspects of cybersecurity. You gained knowledge about systems security and endpoint protection, highlighting antivirus, antimalware, firewalls, and IDS/IPS. We also covered application security, emphasizing secure software development, code reviews, static analysis, and hardening authentication and authorization. Key concepts such as data encryption in transit and at rest and session management were discussed. The chapter concluded with an emphasis on the importance of security patching and updates for a robust cybersecurity strategy. The next chapter will pivot to penetration testing, discussing its methodologies and significance in identifying and addressing security vulnerabilities.

In the next chapter, we will unfold the final part of protecting critical infrastructure, providing an overview of IR, security culture and awareness, and executive orders. This chapter aims to equip you with advanced knowledge and strategies for effectively responding to cybersecurity incidents, fostering a security-conscious culture within organizations, and understanding the impact of executive orders on cybersecurity practices. These areas are essential for a comprehensive approach to safeguarding critical infrastructures against evolving cyber threats.

References

To learn more about the topics covered in this chapter, take a look at the following resources:

- *Sobeslav, V., Balik, L., Hornig, O., Horalek, J., & Krejcar, O. (2017). Endpoint firewall for local security hardening in academic research environment. Journal of Intelligent & Fuzzy Systems, 32(2), 1475-1484.*

- *Pagán, A., & Elleithy, K. (2021, January). A multi-layered defense approach to safeguard against ransomware. In 2021 IEEE 11th Annual Computing and Communication Workshop and Conference (CCWC) (pp. 0942-0947). IEEE.*

- *Solanki, P. S., Kadam, A. D., Khairnar, P. S., & Atkekar, N. D. (2022). Enhancement of cyber security and protection of sensitive R&D data using next generation firewall.*

- *Min, B., & Varadharajan, V. (2016). A novel malware for subversion of self-protection in anti-virus. Software: Practice and Experience, 46(3), 361-379.*

- *Morales, J., Xu, S., & Sandhu, R. (2012). Analyzing malware detection efficiency with multiple anti-malware programs. ASE Science Journal, 1(2), 56-66.*

- *Cabrera, E. (2016). Protecting critical infrastructure from cyberattack. Risk Management, 63(8), 32.*

- *Nash, T. (2005). An undirected attack against critical infrastructure. Technical Report, US-CERT Control Systems Security Center.*

- *Makrakis, G. M., Kolias, C., Kambourakis, G., Rieger, C., & Benjamin, J. (2021). Industrial and critical infrastructure security: Technical analysis of real-life security incidents. Ieee Access, 9, 165295-165325.*

- *Sandaruwan, G. P. H., Ranaweera, P. S., & Oleshchuk, V. A. (2013, December). PLC security and critical infrastructure protection. In 2013 IEEE 8th international conference on industrial and information systems (pp. 81-85). IEEE.*

- *Scott, J. (2017). Signature based malware detection is dead. Institute for Critical Infrastructure Technology.*

- *Rege, A., & Bleiman, R. (2020, June). Ransomware attacks against critical infrastructure. In Proc. 20th Eur. Conf. Cyber Warfare Security (p. 324).*

- *Ibarra, J., Butt, U. J., Do, A., Jahankhani, H., & Jamal, A. (2019, January). Ransomware impact to SCADA systems and its scope to critical infrastructure. In 2019 IEEE 12th International Conference on Global Security, Safety and Sustainability (ICGS3) (pp. 1-12). IEEE.*

- *Heino, J., Hakkala, A., & Virtanen, S. (2022). Study of methods for endpoint aware inspection in a next generation firewall. Cybersecurity, 5(1), 25.*

- *Sharma, R., & Parekh, C. (2017). Firewalls: A Study and Its Classification. International Journal of Advanced Research in Computer Science, 8(5).*

- *Wack, J., Cutler, K., & Pole, J. (2002). Guidelines on firewalls and firewall policy. NIST special publication, 800, 41.*

- *Hollis, S., & Zahn, D. (2017). ICS Cybersecurity: Protecting the Industrial Endpoints That Matter Most.*

- *Ginter, A. F. (2017). Cyber Perimeters for Critical Infrastructures. Cyber-Physical Security: Protecting Critical Infrastructure at the State and Local Level, 67-100.*

- *Makrakis, G. M., Kolias, C., Kambourakis, G., Rieger, C., & Benjamin, J. (2021). Industrial and critical infrastructure security: Technical analysis of real-life security incidents. Ieee Access, 9, 165295-165325.*

- *Letou, K., Devi, D., & Singh, Y. J. (2013). Host-based intrusion detection and prevention system (HIDPS). International Journal of Computer Applications, 69(26), 28-33.*

- *Scarfone, K., & Mell, P. (2007). Guide to intrusion detection and prevention systems (idps). NIST special publication, 800(2007), 94.*

- *Liu, M., Xue, Z., Xu, X., Zhong, C., & Chen, J. (2018). Host-based intrusion detection system with system calls: Review and future trends. ACM computing surveys (CSUR), 51(5), 1-36.*

- Koller, R., Rangaswami, R., Marrero, J., Hernandez, I., Smith, G., Barsilai, M., ... & Merrill, K. (2008, June). Anatomy of a real-time intrusion prevention system. In 2008 International Conference on Autonomic Computing (pp. 151-160). IEEE.

- Sun, M., Zheng, M., Lui, J. C., & Jiang, X. (2014, December). Design and implementation of an android host-based intrusion prevention system. In Proceedings of the 30th annual computer security applications conference (pp. 226-235).

- Hassan, W. U., Bates, A., & Marino, D. (2020, May). Tactical provenance analysis for endpoint detection and response systems. In 2020 IEEE Symposium on Security and Privacy (SP) (pp. 1172-1189). IEEE.

- Agarwal, S., Sable, A., Sawant, D., Kahalekar, S., & Hanawal, M. K. (2022, January). Threat detection and response in Linux endpoints. In 2022 14th International Conference on COMmunication Systems & NETworkS (COMSNETS) (pp. 447-449). IEEE.

- Chandel, S., Yu, S., Yitian, T., Zhili, Z., & Yusheng, H. (2019, October). Endpoint protection: Measuring the effectiveness of remediation technologies and methodologies for insider threat. In 2019 international conference on cyber-enabled distributed computing and knowledge discovery (cyberc) (pp. 81-89). IEEE.

- Futcher, L., & Von Solms, R. (2008, October). Guidelines for secure software development. In Proceedings of the 2008 annual research conference of the South African Institute of Computer Scientists and Information Technologists on IT research in developing countries: riding the wave of technology (pp. 56-65).

- Roshaidie, M. D., Liang, W. P. H., Jun, C. G. K., & Yew, K. H. (2020). Importance of Secure Software Development Processes and Tools for Developers. arXiv preprint arXiv:2012.15153.

- Ayewah, N., Pugh, W., Hovemeyer, D., Morgenthaler, J. D., & Penix, J. (2008). Using static analysis to find bugs. IEEE software, 25(5), 22-29.

- Livshits, V. B., & Lam, M. S. (2005, August). Finding Security Vulnerabilities in Java Applications with Static Analysis. In USENIX security symposium (Vol. 14, pp. 18-18).

- Aradau, C. (2010). Security that matters: Critical infrastructure and objects of protection. Security dialogue, 41(5), 491-514.

- Jang-Jaccard, J., & Nepal, S. (2014). A survey of emerging threats in cybersecurity. Journal of computer and system sciences, 80(5), 973-993.

- Case, D. U. (2016). Analysis of the cyber attack on the Ukrainian power grid. Electricity Information Sharing and Analysis Center (E-ISAC), 388(1-29), 3.

- Kimani, K., Oduol, V., & Langat, K. (2019). Cyber security challenges for IoT-based smart grid networks. International journal of critical infrastructure protection, 25, 36-49.

- Igure, V. M., Laughter, S. A., & Williams, R. D. (2006). Security issues in SCADA networks. computers & security, 25(7), 498-506.

- *Alenezi, M. N., Alabdulrazzaq, H., & Mohammad, N. Q. (2020). Symmetric encryption algorithms: Review and evaluation study. International Journal of Communication Networks and Information Security, 12(2), 256-272.*

- *Alenezi, M. N., Alabdulrazzaq, H., & Mohammad, N. Q. (2020). Symmetric encryption algorithms: Review and evaluation study. International Journal of Communication Networks and Information Security, 12(2), 256-272.*

- *Trevisan, M., Soro, F., Mellia, M., Drago, I., & Morla, R. (2020). Does domain name encryption increase users' privacy? ACM SIGCOMM Computer Communication Review, 50(3), 16-22.*

- *Malatji, M., Marnewick, A. L., & Von Solms, S. (2022). Cybersecurity capabilities for critical infrastructure resilience. Information & Computer Security, 30(2), 255-279.*

- *Seth, B., Dalal, S., Jaglan, V., Le, D. N., Mohan, S., & Srivastava, G. (2022). Integrating encryption techniques for secure data storage in the cloud. Transactions on Emerging Telecommunications Technologies, 33(4), e4108.*

- *Resul, D. A. S., & Gündüz, M. Z. (2020). Analysis of cyber-attacks in IoT-based critical infrastructures. International Journal of Information Security Science, 8(4), 122-133.*

- *Dissanayake, N., Jayatilaka, A., Zahedi, M., & Babar, M. A. (2022). Software security patch management-A systematic literature review of challenges, approaches, tools and practices. Information and Software Technology, 144, 106771.*

- *Serror, M., Hack, S., Henze, M., Schuba, M., & Wehrle, K. (2020). Challenges and opportunities in securing the industrial internet of things. IEEE Transactions on Industrial Informatics, 17(5), 2985-2996.*

- *Yaacoub, J. P. A., Noura, H. N., Salman, O., & Chehab, A. (2023). Ethical hacking for IoT: Security issues, challenges, solutions and recommendations. Internet of Things and Cyber-Physical Systems, 3, 280-308.*

- *Tawalbeh, L. A., Muheidat, F., Tawalbeh, M., & Quwaider, M. (2020). IoT Privacy and security: Challenges and solutions. Applied Sciences, 10(12), 4102.*

- *Faquir, D., Chouliaras, N., Sofia, V., Olga, K., & Maglaras, L. (2021). Cybersecurity in smart grids, challenges and solutions. AIMS Electronics and Electrical Engineering, 5(1), 24-37.*

- *Cavusoglu, H., Cavusoglu, H., & Zhang, J. (2008). Security patch management: Share the burden or share the damage? Management Science, 54(4), 657-670.*

- *Olswang, A., Gonda, T., Puzis, R., Shani, G., Shapira, B., & Tractinsky, N. (2022). Prioritizing vulnerability patches in large networks. Expert Systems with Applications, 193, 116467.*

8
Protecting Critical Infrastructure – Part 3

In the preceding chapters of this comprehensive exploration into the safeguarding of critical infrastructure, we've journeyed through the intricate layers of defense, meticulously fortifying the foundation upon which the resilience of our nation's lifelines rests. We've discussed systems security and endpoint protection, dived into the complexities of application security, and delved into the nuances of network security, continuous monitoring, and the establishment of policy and security frameworks.

As we continue with this new chapter, we transition from the realm of proactive prevention to the dynamic and critical aspects of **incident response** (**IR**), security culture and awareness, and the instrumental role of security executive orders in preserving our critical infrastructure.

In an age where threats to our lifelines evolve at an unprecedented pace, the ability to respond swiftly and effectively to security incidents is paramount. This chapter will guide us through the strategies and tactics employed to detect, contain, and recover from incidents, ensuring that our critical infrastructure remains robust and resilient in the face of adversity.

But protecting our nation's lifelines goes beyond technology and procedures. It extends deep into the realm of human behavior and culture. We will emphasize the significance of cultivating a culture of security awareness within organizations and communities. This culture fosters a collective commitment to vigilance and preparedness, making it an indispensable pillar of critical infrastructure defense.

Furthermore, we will delve into the realm of government leadership and its profound impact on fortifying our lifelines. Security executive orders serve as powerful tools in shaping the landscape of critical infrastructure protection. We will examine how these directives influence security measures, exploring the complexities and challenges inherent in their implementation.

Together, the topics of IR, security culture and awareness, and security executive orders build upon the strong foundations laid in previous chapters, ensuring the security, resilience, and preparedness of our nation's critical infrastructure.

In this chapter, we will cover the following topics:

- Incident response (IR)
- Security culture and awareness
- Executive orders

Incident response (IR)

This section explores the evolution of IR in the context of cybersecurity and computer security. It traces its origins to the early days of computer technology and the internet, highlighting the increasing importance of IR as computer systems became integral to business and government operations.

Key historical figures, such as Clifford Stoll, are discussed for their significant contributions to IR in the late 1980s at **Lawrence Berkeley National Laboratory (LBNL)**.

This section also discusses the establishment of **computer emergency response teams (CERTs)** and the development of industry standards and best practices in IR, such as the **National Institute of Standards and Technology (NIST)** *Computer Security Incident Handling Guide*.

This section underscores the continuous evolution of IR in addressing new types of cyber threats and emphasizes the need for effective IR planning.

IR history

IR, in the context of cybersecurity and computer security, originated as a field of practice and study in response to growing threats and vulnerabilities associated with the proliferation of computer technology and the internet.

The need for IR can be traced back to the early days of computer technology when organizations first started using computers for business and government purposes. As computer systems became more critical to operations, concerns about security and unauthorized access emerged.

In the 1980s and 1990s, the internet began to grow, and with it came the emergence of hacking and malware. Incidents of computer breaches, data theft, and disruptive attacks started to become more common. This led to a growing need for organizations to develop strategies to respond to these incidents.

Clifford Stoll homage

Clifford Stoll is a well-known American astronomer, author, and computer security expert who played a significant role in an incident at LBNL in the late 1980s. His story became widely known through his book *The Cuckoo's Egg: Tracking a Spy Through the Maze of Computer Espionage*, which detailed his experiences at LBNL and his pursuit of a hacker.

In 1986, Clifford Stoll was working as a systems administrator at LBNL, primarily dealing with computer systems used for astrophysical research. One day, he noticed a 75-cent accounting discrepancy in the lab's computer network. Initially dismissing it as a minor issue, Stoll decided to investigate further out of curiosity.

As Stoll delved deeper into the seemingly insignificant discrepancy, he discovered that it was caused by a hacker who had infiltrated the lab's computer systems. Over several months, Stoll meticulously tracked the hacker's activities, following a trail of log files, network connections, and suspicious behavior.

Stoll's investigation eventually revealed that the hacker, who went by the pseudonym *Hunter*, was not just an ordinary hacker but likely a foreign spy working for the KGB, the Soviet Union's intelligence agency. Hunter had been using LBNL's computer network as a gateway to gain unauthorized access to various military and research institutions across the United States.

The Cuckoo's Egg chronicles Stoll's pursuit of the hacker, his interactions with law enforcement agencies, and his efforts to uncover the extent of the espionage operation. His story highlighted the importance of computer security and the potential consequences of cyberattacks, even in the early days of the internet.

Stoll's work and the subsequent book helped raise awareness about cybersecurity threats and the need for better security practices. It also showcased the value of vigilant individuals such as Stoll in identifying and responding to security incidents.

Clifford Stoll's story at LBNL is often cited as one of the early instances of a cybersecurity IR and investigation that ultimately led to the apprehension of a hacker involved in espionage. His work has had a lasting impact on the field of computer security and has inspired others to take cybersecurity seriously.

CERTs were some of the earliest formal IR organizations. The first CERT, the **CERT Coordination Center (CERT/CC)**, was established by the U.S. Department of Defense in 1988 at Carnegie Mellon University. CERTs were created to help organizations respond to and mitigate security incidents.

As the importance of IR became more evident, various industry standards and best practices began to emerge. One of the most influential documents in this regard is *Computer Security Incident Handling Guide*, published by NIST in the United States. This guide provided a framework for organizations to establish IR capabilities.

Governments and regulatory bodies around the world started to recognize the need for cybersecurity regulations and guidelines. Laws such as the **Health Insurance Portability and Accountability Act (HIPAA)**, the Gramm-Leach-Bliley Act, and the Sarbanes-Oxley Act in the United States, as well as the European Union's **General Data Protection Regulation (GDPR)**, require organizations to have **IR plans (IRPs)** in place.

As technology continued to evolve, so did the nature of cyber threats. **IR teams (IRTs)** had to adapt to new types of attacks, including **advanced persistent threats (APTs)**, ransomware, and nation-state-sponsored cyberattacks. This required continuous improvement and adaptation of IR strategies and tools.

The sharing of **threat intelligence (TI)** among organizations and government agencies became an essential part of IR. Information sharing allows organizations to proactively defend against known threats and respond more effectively to incidents.

Various IR frameworks, such as the *NIST Cybersecurity Framework*, the *ISO/IEC 27035* standard, and the SANS Institute's *Incident Handler's Handbook*, were developed to provide guidelines and methodologies for building and executing IRPs.

Today, IR is a critical component of cybersecurity, with organizations of all sizes and types investing in IRTs, processes, and technologies to detect, respond to, and recover from security incidents effectively. The field continues to evolve as new threats and challenges emerge in the ever-changing landscape of cybersecurity.

IR planning

An IRP is a structured and documented set of procedures and guidelines that an organization follows when it encounters a security incident or data breach. The primary purpose of an IRP is to outline steps and actions that need to be taken to identify, manage, mitigate, and recover from a security incident effectively and efficiently.

Key components of an IRP typically include the following:

IR Component	Description
Incident identification	How incidents are detected or reported within the organization, including the responsible parties and methods
Incident classification and severity levels	A categorization of incidents based on their impact, severity, and potential harm to the organization
IRT	Identification of individuals or teams responsible for responding to different types of incidents, their roles, and contact information
Incident investigation and analysis	Procedures for collecting, preserving, and analyzing evidence related to the incident, including forensic analysis if necessary

IR Component	Description
Communication plan	How and when to communicate with internal stakeholders, external authorities, customers, and the public while ensuring compliance with legal and regulatory requirements
Containment and eradication	Actions to limit the damage caused by the incident, remove the threat, and prevent it from spreading further
Recovery and restoration	Steps to restore affected systems and services to normal operation and ensure **business continuity (BC)**
Legal and regulatory compliance	Ensuring that all actions taken during IR align with relevant laws and regulations, including data breach notification requirements
Post-incident review	An assessment of the IR process to identify areas for improvement and lessons learned, which can inform future updates to the plan
Training and awareness	Ongoing education and training for employees so that they are aware of their roles and responsibilities during an incident
Documentation and reporting	Detailed records of the incident, response actions taken, and any findings, which may be needed for legal and compliance purposes

Table 8.1 – Key components of an IRP

Establishing a comprehensive IRP is vital for effectively minimizing the repercussions of security breaches, safeguarding confidential information, and preserving stakeholder confidence. This plan ensures swift and efficient action to lessen the damage in the event of a security incident.

Security culture and awareness

Security culture and awareness are crucial, and often overlooked, aspects of cybersecurity. They encompass attitudes, behaviors, and practices of individuals and organizations when it comes to safeguarding digital assets and information. It refers to the overall mindset and values of an organization regarding cybersecurity. It encompasses how seriously an organization takes security and how ingrained security practices are in its daily operations.

Security awareness focuses on educating individuals within an organization about cybersecurity threats, best practices, and their role in protecting sensitive data. Security awareness programs aim to ensure that employees are informed about potential risks, such as phishing attacks, malware, and social engineering tactics. These programs often involve training sessions, workshops, simulated attacks, and ongoing communication to keep employees vigilant and informed.

A strong security culture encourages all employees, from top management to entry-level staff, to prioritize security as an integral part of their roles and responsibilities. Elements of a strong security culture include leadership support, clear security policies and procedures, accountability, and a commitment to ongoing education and improvement.

The following table breaks down the importance of security culture and awareness of cybersecurity into key concepts:

Concept	Description
Prevention	A strong security culture and awareness can help prevent security breaches and incidents by ensuring that all employees are vigilant and follow best practices.
Detection and response	In the event of a security incident, an organization with a robust security culture will be better prepared to detect the breach early and respond effectively.
Compliance	Many industries have regulatory requirements that mandate security awareness and training programs. Maintaining a security-conscious culture helps ensure compliance with these regulations.
Reputation	A breach can damage an organization's reputation. Demonstrating a commitment to cybersecurity through a strong culture and awareness efforts can help maintain trust with customers and stakeholders.
Cost reduction	Preventing security incidents is typically more cost-effective than dealing with the aftermath of a breach, including potential legal and financial consequences.
Continuous improvement	Security culture and awareness efforts should evolve to adapt to new threats and technologies, ensuring that an organization stays ahead of emerging risks.

Table 8.2 – Key concepts of security culture and awareness

Security culture and awareness are very important in the context of critical infrastructure, as these sectors play a vital role in the functioning of a nation's economy, security, and well-being. Let's see some example scenarios where security culture plays a critical role.

Interconnectivity of critical infrastructure

Critical infrastructure systems are often interconnected and rely heavily on digital technology. This interconnectivity creates a larger attack surface, making them attractive targets for cyberattacks. Consider, for example, a modern urban transportation system relying heavily on digital technology and interconnected components, including public transportation nodes, traffic management, payment systems, and centralized control centers. Digital technology optimizes traffic, enhances passenger convenience, and ensures safety, but it introduces vulnerabilities such as the following:

- Cyberattacks on centralized control centers can disrupt services, leading to delays and chaos
- Manipulating real-time information can mislead passengers
- Manipulating payment systems can lead to financial fraud
- Compromised control systems may pose safety risks, such as unauthorized access to tracks or traffic signal manipulation

A strong security culture ensures employees understand their role in cybersecurity, fostering responsibility for safeguarding digital infrastructure. Regular awareness programs empower them to recognize and respond to potential threats effectively, ultimately enhancing the security and resilience of the transportation system.

Cascading effects of a cyberattack

A cyberattack on one component of critical infrastructure can have cascading effects that disrupt multiple sectors. For example, a cyberattack on a power grid can impact water treatment facilities, transportation systems, and healthcare services. A security-conscious culture can help prevent such attacks, and awareness can facilitate a coordinated response.

Imagine a cyberattack that targets a nation's power grid, causing a massive blackout. This single breach has cascading effects:

- **Water facilities**: Water treatment plants lose power, disrupting the water supply
- **Transportation**: Traffic lights and transportation systems stop, causing congestion and safety risks
- **Healthcare**: Hospitals face equipment failures, delaying patient care

A strong security culture in each sector would have minimized vulnerabilities and encouraged early detection of the attack. Cybersecurity awareness and coordination among sectors would have facilitated a quicker, more effective response to mitigate the impact on society.

Responsibility to safeguard critical assets

Protecting critical infrastructure is an issue of public safety and national security. Hostile actors, including nation-states, may attempt to infiltrate and disrupt these systems. A strong security culture instills a sense of responsibility among personnel to safeguard critical assets.

Imagine a nation's telecommunications network, crucial for citizen communication, government operations, and economic stability. Hostile nation-state actors launch a cyberattack to infiltrate and disrupt this critical infrastructure.

Protecting the telecommunications network is essential for national security due to its role in information flow, government operations, and the economy.

A strong security culture within the telecommunications sector is critical, instilling responsibility among employees to safeguard the network and facilitating proactive cybersecurity practices to prevent unauthorized access and respond effectively to cyber threats.

Insider threats

Insider threats are a significant concern in critical infrastructure. Malicious or negligent employees can pose substantial risks. Security awareness programs can help employees recognize and report suspicious behavior, contributing to early threat detection.

In a critical infrastructure setting, such as a nuclear power plant, insider threats can come from employees. An employee might intentionally harm the plant's operations, such as manipulating safety protocols or compromising security.

Another scenario involves well-intentioned employees making mistakes that lead to cybersecurity breaches, such as opening malicious email attachments.

To address these threats, security awareness programs train employees to recognize suspicious behavior, report concerns, and foster a culture of collective responsibility for infrastructure security. These programs help detect and mitigate threats early, enhancing critical infrastructure security and resilience.

Teamwork and information sharing

Critical infrastructure protection often requires collaboration between various government agencies, private-sector organizations, and cybersecurity experts. A strong security culture promotes teamwork and information sharing, while security awareness programs ensure that all stakeholders understand their roles and responsibilities.

Think of a major international airport, a critical infrastructure asset. Protecting it involves collaboration:

- **Government agencies: Transport Security Administration (TSA), Federal Aviation Administration (FAA)**, and law enforcement create security regulations and oversee measures

- **Private-sector organizations**: Airlines, airport operators, and security companies implement security protocols

- **Cybersecurity practitioners**: Specialists continuously assess digital infrastructure looking for vulnerabilities

Within this collaborative environment, a strong security culture is essential. It promotes a sense of shared responsibility and teamwork among all stakeholders, emphasizing the importance of vigilance and cooperation in safeguarding the airport.

Security awareness programs ensure that all individuals involved in airport security understand their roles and responsibilities. Airport staff, security personnel, airline employees, and even passengers are educated about security procedures, threat recognition, and the significance of reporting suspicious activities.

In this example, collaboration, a strong security culture, and security awareness are crucial for safeguarding the airport, a critical infrastructure asset.

A security culture can drive investment in advanced cybersecurity technologies and solutions. However, it's crucial to ensure that employees are also aware of how to use these tools effectively to protect critical infrastructure assets.

The threat landscape in cybersecurity is constantly evolving, with attackers developing new techniques and tactics. A security culture that emphasizes adaptability and continual improvement is essential to stay ahead of emerging risks and vulnerabilities.

Many governments have established regulations and standards for the cybersecurity of critical infrastructure. A robust security culture ensures compliance with these regulations, while security awareness programs help employees understand and adhere to specific requirements.

Let's explore some of the executive orders around security in critical infrastructure in the next section.

Executive orders

Executive orders around security in critical infrastructure are a significant aspect of national policy, particularly in countries such as the United States. These orders are directives issued by a country's leader, such as the President of the United States, to manage the operations of the federal government. They are instrumental in shaping policies, especially in areas where legislative action is slow or contentious.

The increasing sophistication of cyber threats and the potential for terrorist attacks make the security of these critical infrastructures a national security priority. Disruptions in critical infrastructure can also have significant economic consequences, affecting everything from stock markets to the daily lives of citizens.

Many executive orders focus on protecting infrastructure requiring regular assessments of cybersecurity risks and the implementation of robust security measures. These executive orders typically stress the importance of cooperation between government entities and the private sector, given the private ownership of much critical infrastructure. They often mandate enhanced exchanges of information about cyber threats and vulnerabilities, involving both governmental bodies and private sector entities. They also address the need for contingency planning and response strategies for potential physical attacks or cyberattacks.

Over the years, various administrations in the United States have issued executive orders to address evolving threats and technological advancements. For example, post-9/11, there was an increased focus on protecting infrastructure from terrorist attacks.

Executive orders must balance the need for security with the protection of individual privacy and civil liberties. The effectiveness of these orders can be limited by challenges in implementation, especially in coordinating across different sectors and levels of government.

Let's touch base on some of the last two decades' executive orders and presidential directives chronologically.

Executive Order 13010 – Critical Infrastructure Protection (1996)

Issued by President Bill Clinton, this order was a pioneering step in recognizing the importance of protecting critical infrastructures such as telecommunications and electrical power systems from physical and cyber threats. It established the **President's Commission on Critical Infrastructure Protection (PCCIP)**, which was tasked with assessing vulnerabilities and recommending a national strategy.

Key points of *Executive Order 13010* include the following:

- Identifies critical national infrastructures whose incapacity/destruction would impact United States defense or economy
- Categorizes threats into physical and cyber threats
- Calls for a cooperative government and private sector strategy development
- Establishes the PCCIP
- Sets a mission for the commission to develop a national policy and implementation strategy
- Recommends legal and policy changes for infrastructure protection
- Instructs the creation of an advisory committee from the private sector for the commission
- Establishes an interim Infrastructure Protection Task Force within the Department of Justice

Executive Order 13231 – Critical Infrastructure Protection in the Information Age (2001)

Signed by President George W. Bush following the 9/11 attacks, this order emphasized the protection of information systems for critical infrastructure. It established the President's **Critical Infrastructure Protection Board** (**CIPB**) and highlighted growing concerns about cyber threats in the digital age.

Key points of *Executive Order 13231* include the following:

- Creates the President's CIPB to coordinate cybersecurity efforts

- Encourages partnerships between the federal government and the private sector

- Mandates regular assessments of cybersecurity risks and the development of protective measures

- Promotes the sharing of cybersecurity threat and response information among government entities and the private sector

- Aims to improve the federal government's ability to respond to cybersecurity incidents affecting critical infrastructure

- Directs the implementation of programs to protect information systems supporting critical infrastructure sectors

Homeland Security Presidential Directive 7 (HSPD-7) – Critical Infrastructure Identification, Prioritization, and Protection (2003)

While not an executive order, *HSPD-7*, issued by President George W. Bush, holds considerable importance in the realm of protecting critical infrastructure. The policy sets a national directive for federal departments and agencies to pinpoint and rank critical infrastructure and key resources of the United States, with the aim of safeguarding them against terrorist threats.

Key points of *HSPD-7* include the following:

- Sets forth a national directive for federal departments and agencies to recognize and classify critical infrastructure and essential resources in the United States for defense against acts of terrorism

- Seeks to avert, discourage, and lessen the impact of assaults on vital infrastructure that may result in extensive casualties, undermine economic stability, or harm public confidence

- Defines terms such as *critical infrastructure* and *key resources* and outlines the roles of various federal, state, and local agencies

- Advocates for the creation of an all-encompassing, cohesive national strategy for the safeguarding of critical infrastructure and key resources

- Mandates that federal departments and agencies collaborate with private entities to establish and implement security measures for infrastructure

- Stipulates annual reporting by sector-specific agencies on the progress of infrastructure protection efforts

Executive Order 13636 – Improving Critical Infrastructure Cybersecurity (2013)

Issued by President Barack Obama, this order called for the development of a voluntary risk-based cybersecurity framework – a set of industry standards and best practices to help organizations manage cybersecurity risks. Although it's from 2013, it's worth mentioning as it laid the groundwork for many subsequent initiatives in cybersecurity.

Key points of *Executive Order 13636* include the following:

- Development of a cybersecurity framework by NIST, incorporating industry best practices and standards

- Promotion of information sharing between the government and the private sector regarding cyber threats and security measures

- Ensuring the protection of civil liberties and privacy in the implementation of cybersecurity measures

- Establishment of a voluntary program to support the adoption of the NIST Cybersecurity Framework by critical infrastructure entities

- Identification and prioritization of critical infrastructure at greatest risk from cyber threats, with annual reviews

Presidential Policy Directive 21 (PPD-21) – Critical Infrastructure Security and Resilience (2013)

While not an executive order, this directive from President Barack Obama was pivotal. It aimed to strengthen and maintain secure, functioning, and resilient critical infrastructure. *PPD-21* identified 16 critical infrastructure sectors and emphasized an integrated, collaborative approach between the government and the private sector.

Key points of *PPD-21* include the following:

- Establishing a national unity of effort for secure, functional, and resilient critical infrastructure

- Emphasizing the diverse and complex nature of infrastructure, requiring collaborative risk management efforts

- Mandating integration with the national preparedness system across prevention, protection, mitigation, response, and recovery

- Assigning specific roles and responsibilities to federal departments and agencies while promoting partnerships with critical infrastructure owners and operators

- Outlining strategies to address physical and cyber threats, vulnerabilities, and interdependencies

Executive Order 13873 – Securing the Information and Communications Technology and Services Supply Chain (2019)

Signed by President Donald Trump, this order prohibited transactions posing an undue risk of sabotage to the United States' information and communications technology and services supply chain. It was a significant step in protecting national security against foreign adversaries.

Key points of *Executive Order 13873* include the following:

- Prohibition of transactions that pose risks to the United States' critical infrastructure or digital economy

- Authority granted to the Secretary of Commerce to prohibit transactions involving foreign adversaries

- Development of a framework to safeguard against foreign exploitation of ICT vulnerabilities

- Requirement for recurring and final reports to Congress on the national emergency regarding ICT supply chain threats

- Assessment of threats from foreign entities and the identification of vulnerable entities and products within the United States

Executive Order 13870 – America's Cybersecurity Workforce (2019)

President Trump issued this order to enhance national cybersecurity by improving the federal government's cybersecurity workforce. While not directly targeting infrastructure, it addressed a key aspect of cybersecurity – human resources and skills.

Executive Order 13870 aims to do the following:

- Mobilize resources to address cybersecurity workforce needs

- Enhance the cybersecurity learning environment for a skilled workforce

- Align education and training with employer needs for lifelong cybersecurity careers

- Use measures to assess the effectiveness of cybersecurity workforce investments

Executive Order 13865 – Coordinating National Resilience to Electromagnetic Pulses (2019)

This order, also by President Trump, recognized the threat of **electromagnetic pulses** (**EMPs**) to critical infrastructure, such as the power grid. It aimed to enhance national resilience to EMPs through research, risk assessments, and mitigation strategies.

Executive Order 13865 focuses on enhancing national resilience against EMP and **geomagnetic disturbance** (**GMD**) events. Here are the key points summarized:

- The **Department of Homeland Security** (**DHS**) is charged with coordinating national resilience, preparedness, and response efforts for EMP and GMD events
- The order directs DHS, alongside other federal agencies, to coordinate actions to mitigate the effects of EMPs and GMDs, including extreme space weather events
- The **Cybersecurity and Infrastructure Security Agency** (**CISA**) leads the DHS effort to improve risk awareness and enhance protective measures for critical infrastructure against electromagnetic threats
- Promotes effective electromagnetic IR and recovery activities

Executive Order 13905 – Strengthening National Resilience through Responsible Use of Positioning, Navigation, and Timing Services (2020)

Issued by President Trump, this order aimed to reduce the vulnerability of critical infrastructure to disruptions in GPS services. It sought to ensure the responsible use of these services for national security and economic vitality.

Executive Order 13905 focuses on the following:

- Ensuring the resilience of critical infrastructure by fostering responsible use of **positioning, navigation, and timing** (**PNT**) services
- Defining PNT services and setting standards for their use to minimize risk
- Identifying systems, networks, and assets that depend on PNT services and managing associated risks
- Developing PNT profiles for public and private sectors to follow
- Testing vulnerabilities in critical infrastructure against PNT disruptions
- Creating plans to engage with infrastructure owners/operators for responsible PNT use

- Coordinating R&D for robust and secure PNT services
- Making an independent source of **Coordinated Universal Time** (**UTC**) available for public and private sectors

Executive Order 14028 – Improving the Nation's Cybersecurity (2021)

Issued by President Joe Biden, this order aimed to bolster the nation's cybersecurity in response to significant cyber incidents, including the SolarWinds hack. It focused on modernizing cybersecurity defenses, enhancing software supply chain security, establishing a cybersecurity safety review board, and improving the detection of cybersecurity incidents on federal government networks.

Key points of *Executive Order 14028*, signed on May 12, 2021, include the following:

- Enhancing information sharing between the government and the private sector
- Establishing a standard for software sold to the government, including a **Software Bill of Materials** (**SBOM**)
- Implementing stronger cybersecurity standards within federal agencies, including employing a Zero Trust security model
- Improving detection of cybersecurity incidents on federal networks
- Strengthening the federal government's response to cybersecurity vulnerabilities and incidents
- Improving the federal government's investigative and remediation capabilities

Executive Order 14110 – Safe, Secure, and Trustworthy Development and Use of Artificial Intelligence (2023)

President Biden's **artificial intelligence** (**AI**)-focused executive order is pivotal for ensuring the responsible integration of AI technologies into essential national systems. The directive includes the formulation of safety protocols and rigorous testing of AI to preemptively address any weaknesses. It compels agencies to scrutinize and lessen AI-related security threats to the nation's critical infrastructure, underscoring the commitment to safeguard the core services that underpin the safety and well-being of the public while upholding American principles and the welfare of the workforce.

To summarize, the key points of *Executive Order 14110* are as follows:

- Promotes AI development that is safe, secure, and trustworthy while protecting civil rights and American values
- Ensures AI does not exacerbate inequity or discrimination and upholds privacy and civil liberties
- Aims for responsible use of AI in law enforcement and the prevention of misuse in surveillance and decision-making processes

- Encourages the participation of workers and unions in the development and use of AI

- Focuses on enhancing national AI infrastructure, increasing federal governance, and enabling American leadership in setting global AI standards

These executive orders reflect the evolving nature of threats to critical infrastructure and the United States government's commitment to addressing these challenges through policy directives. These orders demonstrate the evolving understanding and approach of the United States government toward the security of critical infrastructure, spanning from concerns about physical threats to an increasing focus on cybersecurity. The progression of these directives reflects the changing nature of threats and the growing dependence on digital technologies in critical infrastructure sectors.

In conclusion, executive orders on security in critical infrastructure play a crucial role in national security. They are dynamic instruments that must evolve with changing threats and technological advancements. Their success hinges on effective implementation, collaboration across sectors, and balancing security needs with civil liberties.

Summary

This chapter encapsulates the comprehensive journey of safeguarding critical infrastructure, summarizing the layered strategies of defense that have been discussed. It transitions to examining the critical roles of IR, security culture, and the strategic impact of executive orders in reinforcing the resilience of essential services. The chapter acknowledges fast-evolving threats to infrastructure and emphasizes the need for swift and effective incident management to maintain robustness against such challenges.

Key to this protection is fostering a strong security culture and awareness within organizations, which involves creating a collective commitment to security vigilance. We have also critically assessed how executive orders have historically shaped and will continue to influence protective measures for infrastructure, considering the complexities and challenges in their implementation.

Looking ahead, the next chapter will get into the future of AI, exploring its implications for critical infrastructure security, including the risks and opportunities it presents, and envisioning how AI can be developed and utilized in a safe, secure, and trustworthy manner in line with strategic directions set by government policies.

References

To learn more about the topics covered in this chapter, take a look at the following resources:

- *Stoll, Cliff. The Cuckoo's Egg: Tracking a Spy Through the Maze of Computer Espionage. Simon and Schuster, 2005.*

- *Stoll, C. (1988). Stalking the wily hacker. Communications of the ACM, 31(5), 484-497.*

- *Staves, A., Balderstone, H., Green, B., Gouglidis, A., & Hutchison, D. (2020, May). A Framework to Support ICS Cyber Incident Response and Recovery. In ISCRAM (pp. 638-651).*

- Cichonski, P., Millar, T., Grance, T., & Scarfone, K. (2012). *Computer Security Incident Handling Guide (NIST Special Publication 800-61 Rev. 2). National Institute of Standards and Technology*: `https://nvlpubs.nist.gov/nistpubs/SpecialPublications/NIST.SP.800-61r2.pdf`

- Slayton, Rebecca, and Brian Clarke. "Trusting infrastructure: The emergence of computer security incident response, 1989–2005." *Technology and Culture 61.1 (2020): 173-206.*

- Schumaker, Erin. "What is a HIPAA violation?" *ABC News (2021).*

- U.S. Department of Health & Human Services. (n.d.). *HIPAA for professionals: Laws & regulations*: `https://www.hhs.gov/hipaa/for-professionals/privacy/laws-regulations/index.html`

- GDPR-Info. (n.d.). *General Data Protection Regulation (GDPR)*: `https://gdpr-info.eu/`

- GDPR.eu. (n.d.). *General Data Protection Regulation: Compliance and Information*: `https://gdpr.eu/`

- Xie, Y. X., Ji, L. X., Li, L. S., Guo, Z., & Baker, T. (2021). *An adaptive defense mechanism to prevent advanced persistent threats. Connection Science, 33(2), 359-379.*

- Palleti, V. R., Adepu, S., Mishra, V. K., & Mathur, A. (2021). *Cascading effects of cyber-attacks on interconnected critical infrastructure. Cybersecurity, 4, 1-19.*

- Rajkumar, V. S., Ștefanov, A., Presekal, A., Palensky, P., & Torres, J. L. R. (2023). *Cyber attacks on power grids: Causes and propagation of cascading failures. IEEE Access.*

- Kotidis, A., & Schreft, S. (2022). *Cyberattacks and financial stability: Evidence from a natural experiment.*

- Wells, E. M., Boden, M., Tseytlin, I., & Linkov, I. (2022). *Modeling critical infrastructure resilience under compounding threats: A systematic literature review. Progress in disaster science, 15, 100244.*

- Malatji, M., Marnewick, A. L., & Von Solms, S. (2022). *Cybersecurity capabilities for critical infrastructure resilience. Information & Computer Security, 30(2), 255-279.*

- Mohebbi, S., Zhang, Q., Wells, E. C., Zhao, T., Nguyen, H., Li, M., ... & Ou, X. (2020). *Cyber-physical-social interdependencies and organizational resilience: A review of water, transportation, and cyber infrastructure systems and processes. Sustainable Cities and Society, 62, 102327.*

- Wiley, A., McCormac, A., & Calic, D. (2020). *More than the individual: Examining the relationship between culture and Information Security Awareness. Computers & Security, 88, 101640.*

- Alrammah, I., & Ajlouni, A. W. (2021). *A framework and a survey analysis on nuclear security culture at various radiological facilities. Annals of Nuclear Energy, 158, 108294.*

- Hasan, S., Ali, M., Kurnia, S., & Thurasamy, R. (2021). *Evaluating the cyber security readiness of organizations and its influence on performance. Journal of Information Security and Applications, 58, 102726.*

- *Staves, A., Anderson, T., Balderstone, H., Green, B., Gouglidis, A., & Hutchison, D. (2022). A cyber incident response and recovery framework to support operators of industrial control systems. International Journal of Critical Infrastructure Protection, 37, 100505.*

- *Li, Y., & Liu, Q. (2021). A comprehensive review study of cyber-attacks and cyber security; Emerging trends and recent developments. Energy Reports, 7, 8176-8186.*

- *Georgiadou, A., Mouzakitis, S., & Askounis, D. (2022). Detecting insider threat via a cyber-security culture framework. Journal of Computer Information Systems, 62(4), 706-716.*

- *Whitty, M. T. (2021). Developing a conceptual model for insider threat. Journal of Management & Organization, 27(5), 911-929.*

- *Singh, M., Mehtre, B. M., & Sangeetha, S. (2021, May). User behaviour based insider threat detection in critical infrastructures. In 2021 2nd International Conference on Secure Cyber Computing and Communications (ICSCCC) (pp. 489-494). IEEE.*

- *Ghafir, I., Saleem, J., Hammoudeh, M., Faour, H., Prenosil, V., Jaf, S., ... & Baker, T. (2018). Security threats to critical infrastructure: the human factor. The Journal of Supercomputing, 74, 4986-5002.*

- *Knapp, E. D., & Langill, J. T. (2014). Industrial Network Security: Securing critical infrastructure networks for smart grid, SCADA, and other Industrial Control Systems. Syngress.*

- *Petrenj, B., Lettieri, E., & Trucco, P. (2013). Information sharing and collaboration for critical infrastructure resilience–a comprehensive review on barriers and emerging capabilities. International journal of critical infrastructures, 9(4), 304-329.*

- *Bush, G. (2001). Executive Order 13231: Critical infrastructure protection in the information age. The White House, Washington, DC:* `https://www.dhs.gov/xlibrary/assets/executive-order-13231-dated-2001-10-16-initial.pdf`

- *Bush, G. (2003). Homeland Security Presidential Directive/HSPD7. Washington, DC: The White House:* `https://georgewbush-whitehouse.archives.gov/news/releases/2003/12/20031217-5.html`

- *Obama, B. (2013). Executive Order, 13636, Improving Critical Infrastructure Cybersecurity:* `https://obamawhitehouse.archives.gov/the-press-office/2013/02/12/executive-order-improving-critical-infrastructure-cybersecurity`

- *Obama, B. (2013). Critical infrastructure security and resilience. White House:* `https://obamawhitehouse.archives.gov/the-press-office/2013/02/12/presidential-policy-directive-critical-infrastructure-security-and-resil`

- *Trump, D. J. (2019). Executive order on securing the information and communications technology and services supply chain. The White House:* `https://trumpwhitehouse.archives.gov/presidential-actions/executive-order-securing-information-communications-technology-services-supply-chain/`

- *Trump, D. J. (2019). Executive order on America's cybersecurity workforce*: `https://trumpwhitehouse.archives.gov/presidential-actions/executive-order-americas-cybersecurity-workforce/#:~:text=(d)%20The%20United%20States%20Government,our%20national%20and%20economic%20security`

- *Trump, D. J. (2019). Executive order on coordinating national resilience to electromagnetic pulses. The White House. Executive Order*: `https://trumpwhitehouse.archives.gov/presidential-actions/executive-order-coordinating-national-resilience-electromagnetic-pulses/`

- *Trump, D. J. (2020). Executive Order on Strengthening National Resilience through Responsible Use of Positioning, Navigation, and Timing Services. Executive Order, 13905*: `https://trumpwhitehouse.archives.gov/presidential-actions/executive-order-strengthening-national-resilience-responsible-use-positioning-navigation-timing-services/`

- *Biden, J.R. (2021). Executive Order on Improving the Nation's Cybersecurity. The White House*: `https://www.whitehouse.gov/briefing-room/presidential-actions/2021/05/12/executive-order-on-improving-the-nations-cybersecurity/`

- *Biden, J. R. (2023). Executive order on the safe, secure, and trustworthy development and use of artificial intelligence. The White House*: `https://www.whitehouse.gov/briefing-room/presidential-actions/2021/05/12/executive-order-on-improving-the-nations-cybersecurity/`

References

Part 4: What's Next

Part 4 turns our attention toward the horizon of cybersecurity for critical infrastructure. In this forward-looking section, we consider the future, focusing on how current challenges and emerging technologies such as artificial intelligence will shape the resilience of essential systems. This part is dedicated to preparing for what lies ahead, ensuring that our critical infrastructures can stand resilient against the threats of tomorrow.

This part has the following chapter:

- *Chapter 9, The Future of CI*

9

The Future of CI

The future of cybersecurity in critical infrastructure poses significant challenges. This stark assertion, at first glance, might seem like an unduly pessimistic view. However, when we dig into the complexities of the digital age and the rapid evolution of cyber threats, it becomes clear that this concern is not only warranted but also essential for driving innovation and preparedness in the face of growing challenges.

As we embark on this exploration, it is important to remember the landscape in which CI operates today. These systems—the power plants, water treatment facilities, transportation networks, and communication grids—are the lifeblood of modern society. They have evolved from relatively isolated and manual operations to highly interconnected and automated systems, deeply integrated with information technology and the internet. While this evolution has brought about efficiency and innovation, it has also exposed these vital systems to a myriad of cyber threats.

In recent years, the frequency and sophistication of cyberattacks on CI have escalated alarmingly. Malicious hackers, whether independent actors, criminal organizations, or state-sponsored entities, have demonstrated their ability to penetrate these systems, causing disruptions that range from inconvenient to catastrophic. Incidents such as the shutdown of a major fuel pipeline in the United States or the breach of a city's water treatment system have shown that these are not hypothetical scenarios but real and present dangers.

Looking to the future, several factors suggest that the security landscape for CI will become even more challenging. Firstly, the proliferation of **Internet of Things** (**IoT**) devices has expanded the attack surface dramatically. These devices often lack robust security features, making them easy targets for hackers looking to enter broader systems.

Secondly, the increasing complexity of infrastructure systems themselves makes security more challenging. As these systems become more automated, with legacy software still running in them, it becomes harder to predict and guard against all possible vulnerabilities. The interdependence of these systems also means that a breach in one area can have cascading effects throughout the network.

Thirdly, the geopolitical dimension of cyber threats cannot be ignored. In an era where cyber warfare is becoming a tool of statecraft, CI becomes a prime target for foreign adversaries seeking to disrupt or influence another country.

We stand at a crossroads in history, where the rapid advancement of digital technologies, such as **artificial intelligence** (**AI**), IoT, and advanced cybersecurity measures, intersect with ever-present threats to our physical and cyber systems. The infrastructures that sustain our societies—energy grids, transportation networks, water systems, and communication channels—are evolving, becoming smarter and more efficient, but also more complex and interconnected. With this evolution comes increased vulnerability to a range of threats, from cyberattacks to environmental calamities.

This chapter endeavors to provide a comprehensive overview of what the future holds for the security of these vital systems. It addresses emerging trends, such as increased regulatory requirements, the integration of smart technologies, the significance of **public-private partnerships** (**PPPs**), and the growing need for resilience and rapid recovery in the face of disruptions.

As we navigate through these pages, we invite you to reflect on the profound implications of these changes—not just for those directly involved in the security and management of CI, but for society at large. The future of our infrastructure's security is not just a technical issue; it's a cornerstone for our safety, economic stability, and quality of life.

While the future of cybersecurity in CI presents significant challenges, it also offers opportunities for innovation, collaboration, and advancement.

Let's explore some key trends that are poised to significantly influence the future of cybersecurity in CI. This exploration is crucial for understanding evolving challenges and opportunities in protecting essential systems against emerging cyber threats. We aim to uncover insights and forecast developments while providing the reader with essential knowledge for navigating the dynamic landscape of CI cybersecurity.

In this chapter, we will cover the following topics:

- Increment and innovation of cybersecurity measures
- More robust encryption implementation
- Human factor and training
- PPPs
- Resilience and recovery
- Integration of IoT and smart technologies
- Supply chain security
- Advancements in threat detection technologies
- Greater regulatory and compliance requirements
- Cross-sector collaboration

Increment and innovation of cybersecurity measures

The future of CI security is likely to be shaped by several key trends and developments. One of them is the increment of cybersecurity measures.

The increment of cybersecurity measures is a response to the evolving landscape of cyber threats that continue to grow in complexity and frequency. As we land in this new era, the emphasis on advanced cybersecurity measures will become not just a preference but a necessity for the protection and resilience of CIs.

AI and **machine learning** (**ML**) are at the forefront of this transformation. These technologies have the power to revolutionize how we approach threat detection and response. By leveraging AI and ML, security systems could analyze vast amounts of data, identify patterns, and predict potential threats with greater accuracy than ever before. This capability allows for a more proactive approach to security, moving away from traditional reactive methods.

AI-driven systems can adapt and evolve in response to new threats. Unlike static security measures, these systems learn from each interaction, becoming more sophisticated and effective over time. This adaptability is crucial in a landscape where threat actors continuously employ new tactics and strategies.

That said, it's crucial to understand that AI, despite its rapid advancement and increasing integration into cybersecurity, is not yet at a stage where security can be dependent on this technology. This understanding is vital for a balanced and realistic approach to leveraging AI in CI security.

AI technology, at its current state, still requires strict monitoring and should be considered in exploratory and trial phases in many respects. While AI offers significant advantages in terms of speed, scalability, and efficiency, it also presents unique challenges and limitations that must be acknowledged.

One of the primary concerns is the reliability of AI systems. AI models, particularly those based on ML, rely heavily on the data they are trained on. If this data is biased, incomplete, or not representative of real-world scenarios, the AI's decision-making can be flawed, leading to potential security vulnerabilities or even worse scenarios, such as the denial of service in CIs. Therefore, human oversight is essential to ensure that AI systems are making accurate and appropriate decisions.

Another critical aspect is the interpretability of AI decisions. Many advanced AI systems, especially those based on **deep learning** (**DL**), are often considered *black boxes* because their decision-making processes are not easily understandable by humans. This lack of transparency can be problematic in CI security, where understanding the rationale behind a decision or an alert is crucial for effective response and mitigation strategies.

Additionally, AI systems themselves can become targets of cyberattacks. Adversaries may attempt to manipulate the data or the AI model, leading to a new kind of threat known as adversarial attacks. These attacks can subtly alter the input data in ways that are imperceptible to humans but can cause the AI to make incorrect decisions, compromising the security measures in place. Let's look at some examples in the following table:

Type of Adversarial Attack	Description
Evasion attacks	These involve modifying input data to evade detection by an ML model while appearing normal to human observers.
Poisoning attacks	Attackers inject malicious data into the training set, causing the model to make incorrect predictions or classifications.
Model inversion attacks	Here, the attacker aims to invert the model's predictions to reveal sensitive information about the training data.
Model extraction attacks	Attackers use this method to create a replica of the target model by using its output predictions.
Model inference attacks	This type of attack involves the attacker deducing the properties or structure of the model without explicit extraction.
Backdoor attacks	Attackers implant backdoors in a model during training, which can later be triggered to cause incorrect outputs.
Adversarial patch attacks	These attacks use small patches or images to deceive image recognition systems into misclassifying objects.
Transferability attacks	An attack is crafted against one model and then applied to another model, exploiting the transferability of adversarial examples.

Table 9.1 – Adversarial attacks

Moreover, the integration of AI in cybersecurity is still subject to regulatory and ethical considerations. There are ongoing debates and research regarding the ethical use of AI, privacy concerns, and the need for regulatory frameworks to govern the development and deployment of these technologies in sensitive sectors such as CI.

Therefore, while AI represents a powerful tool in the arsenal of cybersecurity defenses, its current stage of development demands a cautious approach. It should be used as a complement to, rather than a replacement for, traditional security measures and human expertise. Organizations must invest not only in the technology itself but also in building the skills and processes needed to effectively manage, monitor, and interpret AI systems. The potential of AI in cybersecurity is immense, but realizing this potential will require a balanced, informed, and careful approach, especially when it comes to the security of our most CIs.

In addition to AI and ML, other emerging technologies such as blockchain and quantum computing are also set to play a role in the future of cybersecurity. Blockchain offers a decentralized and tamper-resistant ledger system, ideal for secure transactions and data integrity. On the other hand, quantum computing presents both an opportunity and a threat – offering unprecedented computing power to enhance security measures while also posing a potential risk to current encryption standards.

More robust encryption implementation

Another key aspect is the implementation of more robust encryption techniques. As data becomes the backbone of our digital economy, ensuring its integrity and confidentiality is a priority. Advanced encryption methods provide a secure foundation for data transmission and storage, safeguarding against unauthorized access and breaches.

In the intricate digital tapestry of the 21st century, data flows like a lifeblood through the veins of our global economy, carrying with it the pulse of financial transactions, personal communications, and sensitive government information. This data, if compromised, can disrupt economies, endanger personal freedoms, and undermine national security. Thus, the role of encryption transcends mere confidentiality; it is a guardian of trust and stability in the digital realm.

As we traverse deeper into the digital age, encryption must evolve to meet the escalating arms race against cyber threats. We are witnessing the rise of quantum-resistant algorithms, designed to withstand the assault of quantum computing capabilities that could render current encryption obsolete. These sophisticated methods, such as lattice-based cryptography, are being tested and slowly integrated to future-proof our security infrastructure.

However, this move toward more advanced encryption techniques is not a silver bullet. With increased complexity comes the need for stronger key management systems and policies. The challenge lies in distributing and storing these keys securely—often regarded as the Achilles' heel of encryption—since a breach in key management can unravel even the most advanced cryptographic defenses.

Encryption's role in privacy and security often puts it at the center of legal and ethical debates. The right to privacy must be balanced with national security interests, leading to contentious discussions on encryption backdoors for law enforcement and intelligence agencies. These conversations will shape the legal landscape in which encryption technologies develop and are deployed, with significant implications for privacy rights and security.

We must also consider the human element. Encryption, no matter how advanced, requires competent and vigilant operators. Training and awareness are paramount, as the best encryption methods are only as effective as the individuals implementing and managing them.

This brings us back to the importance of necessary cultural shifts within organizations, educational pathways for developing cybersecurity talent, and frameworks for operational excellence that can sustain the secure and ethical use of encryption.

Human factor and training

As technology evolves, so does the need for skilled personnel. Continuous training and awareness programs will be essential to keep staff updated on the latest security practices and technologies.

The incessant advance of technology in the field of cybersecurity brings to the forefront a critical, often undervalued element—the human factor. The tools and strategies that defend our CIs from cyber threats are only as effective as the individuals operating them. In this digital arms race, the need for ongoing education and training is not merely beneficial but imperative.

The landscape of cyber threats is one of constant flux, with adversaries continuously developing new methods to exploit vulnerabilities. In response, cybersecurity practices and technologies must adapt with equal speed and ingenuity. This dynamic environment demands a workforce that is not just skilled but perpetually learning, evolving alongside the technologies they wield.

To address this, continuous training and awareness programs emerge as the linchpin in the defense of our critical systems. These programs must extend beyond the mere impartation of knowledge. They must foster a culture of security-mindedness, where vigilance is as habitual as it is deliberate. Staff at all levels must be equipped not only with technical know-how but also with an understanding of the cyber landscape, recognizing the implications of their actions—or inactions—on the broader security posture.

Awareness programs play a crucial role in this educational ecosystem. They serve to illuminate the often invisible web of cyber risks that thread through everyday activities. From phishing scams to sophisticated social engineering tactics, personnel must be adept at identifying and responding to the spectrum of threats they may encounter.

The human factor also pertains to the cultivation of a diverse and inclusive cybersecurity workforce. Diverse teams bring a multiplicity of perspectives to bear on problem-solving, often leading to more innovative and effective security solutions. As such, training programs must also aim to break down barriers, widen participation, and harness the collective intelligence of a varied pool of talent.

The path ahead is complex, with many organizations grappling with the dual challenges of a cybersecurity skills gap and the rapid pace of technological change. But by placing the human factor at the heart of cybersecurity strategies—through continuous training, awareness, and an inclusive approach—we can fortify our defenses with a workforce that is as resilient and adaptable as the systems it protects.

PPPs

Collaborations between government agencies and private sector companies will become more vital. These partnerships will be crucial for sharing **threat intelligence** (**TI**), best practices, and resources.

In the intricate dance of cybersecurity, the steps of public and private entities are becoming increasingly synchronized. The interconnected nature of today's digital infrastructure means that the ripple effects of a cyberattack can quickly spread beyond the initial target, impacting national security, economic stability, and the public's well-being. Consequently, the siloed approaches of the past are giving way to collaborative strategies, with PPPs emerging as a pivotal element in the cybersecurity world.

These partnerships represent a fusion of capabilities, resources, and expertise from both sectors. Government agencies bring to the table the weight of their regulatory powers, access to intelligence, and the ability to coordinate national and international security efforts. Meanwhile, private sector companies contribute innovative technologies, agile management practices, and sector-specific knowledge.

The synergy of PPPs lies in their ability to facilitate the exchange of critical TI. By sharing information on potential and actual cyber threats, both sectors can develop more comprehensive and timely responses. This shared situational awareness becomes a powerful tool, allowing for preemptive actions and coordinated responses to incidents.

PPPs enable the pooling of resources and expertise to address complex cybersecurity challenges. For instance, the development of secure national infrastructure can benefit from cutting-edge technological solutions and cybersecurity research within the private sector. In turn, private companies can leverage the scale and reach of government initiatives to enhance their own security postures and benefit from guidance on regulatory compliance.

Best practices, too, can be disseminated more effectively through PPPs. With the government's endorsement, security standards developed within the private sector can be adopted more widely, raising the overall security baseline. Additionally, PPPs can drive initiatives for workforce development, addressing the cybersecurity skills gap by fostering talent and providing pathways for professional growth.

Fostering effective PPPs is not without its challenges. Issues of trust, data sharing limitations, and divergent objectives can impede collaboration. To overcome these obstacles, mechanisms for building and sustaining trust, such as confidentiality agreements, clear frameworks for collaboration, and joint exercises and simulations must be deployed.

Resilience and recovery

There will be an increased focus on not just preventing attacks but also on ensuring that infrastructure systems can quickly recover and maintain operations during and after an attack.

In the digital age, resilience and recovery have become watchwords in CI security. As much as prevention is the preferred bulwark against cyber threats, the complex and pervasive nature of digital systems means that breaches are, unfortunately, a matter of *when* and not *if*. Therefore, the ability of infrastructure systems to withstand, adapt to, and rapidly recover from disruptions is a key factor.

This shift toward resilience acknowledges the reality that no defense can be impregnable. It is a pragmatic approach that focuses on minimizing impact, maintaining essential functions, and facilitating a swift return to normal operations. Recovery, in this context, is not just about restoring systems to their previous state but also about learning from incidents to emerge stronger and more secure.

Resilience planning begins with a thorough understanding of the critical functions within an infrastructure system and an assessment of which services must be maintained at all costs. From there, strategies such as redundancy, failovers, and modular design can be employed to ensure that these critical functions are robust against various forms of disruption.

Similarly, recovery strategies must be well conceived, practiced, and agile. They often involve the development of comprehensive **incident response plans** (**IRPs**), the establishment of dedicated recovery teams, and the regular testing of backup systems. The goal is to reduce downtime and the associated economic and social costs of disruptions.

Resilience and recovery cannot be static concepts. They must evolve with the threat landscape and the changing nature of infrastructure systems. This evolution includes the adoption of new technologies and practices, such as automated response mechanisms and the use of cloud-based services for redundancy and flexibility.

But technology is only one piece of the resilience puzzle. Human factors, such as leadership, communication, and training, are equally critical. A resilient organization is one that fosters a culture of preparedness, where staff are trained to respond effectively to incidents and where leaders can make rapid, informed decisions under pressure.

Integration of IoT and smart technologies

The integration of IoT devices and smart technologies into CI systems will continue to grow. This will enhance operational efficiency and data collection but will also introduce new vulnerabilities that must be addressed.

As we chart the course of CI into the future, the proliferation of IoT and smart technologies stands out as both a beacon of progress and a potential source of risk. These technologies promise a revolution in how infrastructure systems are managed and operated, imbuing them with the ability to collect, analyze, and act upon data in real time. The potential benefits are immense: from energy grids that can self-balance supply and demand, to water systems that can detect and respond to contamination automatically.

However, this integration also expands the attack surface available to cyber adversaries. Each device, sensor, or smart component that is added to the network represents a potential entry point for malicious actors. The challenge, then, is to embrace the advantages of IoT and smart technologies while mitigating the inherent risks they bring.

Securing a vast and diverse array of connected devices requires a multifaceted approach. Firstly, there must be stringent standards for device security, ensuring that they are not only resistant to attack but also can be updated to respond to new threats as they emerge. This entails a secure-by-design philosophy that prioritizes security and privacy in the manufacturing process rather than as an afterthought.

Additionally, the complexity of IoT ecosystems calls for robust security protocols and architectures that can handle large-scale device management and data encryption. The interconnectivity of these devices means that a breach in one can have cascading effects throughout the system. Thus, network security becomes critical, necessitating advanced encryption, regular patching, and vigilant monitoring.

The human factor also plays a crucial role in the integration of IoT and smart technologies. Training for personnel must not only cover the technical aspects of these technologies but also instill a keen understanding of security practices needed to manage them effectively.

By understanding the complexities and embracing the rigorous security measures required, we can navigate the challenges and capitalize on the vast potential of IoT and smart technologies to create more efficient, responsive, and sustainable CI systems.

Supply chain security

There will be an increased focus on securing the supply chain of CI components against tampering and cyber espionage.

The integrity of the supply chain is the bedrock upon which the security of CI is built. As the components that constitute our energy systems, telecommunications networks, and transportation hubs are sourced from a complex and globalized supply chain, the potential for compromise grows exponentially. The threat is multifaceted—ranging from the insertion of malicious hardware and software to the exploitation of logistical vulnerabilities—and has profound implications for national security and public safety.

Securing the supply chain, therefore, will remain a main concern, necessitating a comprehensive and strategic approach. This will involve greater scrutiny of suppliers, the implementation of rigorous testing protocols for hardware and software, and the establishment of trusted relationships between suppliers and infrastructure operators.

In recognition of these challenges, there will be a concerted effort to develop more robust standards and best practices for supply chain security. This will likely include measures such as third-party audits, certification processes, and the development of secure manufacturing environments to ensure that the components that make up our CIs are free from tampering.

The advent of cyber-espionage activities targeting the supply chain means that cybersecurity can no longer be an afterthought in the procurement process. Cybersecurity criteria will need to be integrated into the selection of suppliers and throughout the product life cycle, from design to disposal. This integration will ensure that products are not only secure upon delivery but remain secure throughout their operational lifespan.

The human element, once again, is crucial. Education and training will play a significant role in raising awareness about the risks associated with supply chain compromise. Personnel involved in procurement, operations, and security must be equipped to recognize signs of tampering and understand the best practices for mitigating risk.

Advancements in threat detection technologies

Development in areas such as anomaly detection, predictive analytics, and automated response systems will play a significant role in identifying and mitigating threats more efficiently.

As the cyber landscape becomes more hostile and intricate, traditional methods of threat detection and response are being outpaced by the sheer volume and sophistication of attacks. In response, the frontier of cyber defense is rapidly advancing, with cutting-edge technologies emerging as the new vanguard. Among these, anomaly detection, predictive analytics, and automated response systems are the linchpins that will transform our ability to preempt and neutralize cyber threats.

Anomaly detection technologies such as **user and entity behavior analytics (UEBA)** are becoming increasingly nuanced and sophisticated, moving beyond simple rule-based systems to incorporate advanced algorithms that can learn and adapt over time. By utilizing ML and AI, these systems are trained to recognize patterns of normal behavior and, crucially, to identify deviations that may indicate a security breach. As they evolve, these technologies will be able to discern even the most subtle anomalies, reducing false positives and allowing security teams to focus on genuine threats.

Predictive analytics takes threat detection a step further by not just identifying current anomalies but also anticipating future threats. By analyzing historical data and current trends, predictive analytics can forecast potential attack vectors and vulnerabilities. This forward-looking approach enables organizations to bolster their defenses proactively, addressing weak points before they can be exploited.

The third pillar, automated response systems (**security orchestration, automation, and response, or SOAR**), represents a paradigm shift in how cyber incidents are managed. Time is of the essence when responding to breaches, and automated systems can react in milliseconds, much faster than any human. These systems can isolate affected networks, apply patches, or change configurations to mitigate the impact of an attack, often before security teams would even be aware of the breach.

However, the adoption of these advanced technologies is not without challenges. Ensuring the accuracy and efficacy of these systems requires vast amounts of data, skilled personnel to manage and interpret the outputs, and continuous refinement of the algorithms. There is also the issue of trust—organizations must have confidence in automated actions taken on their behalf, understanding that these systems can make critical decisions that have far-reaching consequences.

Greater regulatory and compliance requirements

Governments and international bodies are likely to introduce more stringent regulations and standards (the **Health Insurance Portability and Accountability Act (HIPPA)**, the **General Data Protection Regulation (GDPR)**, the **Federal Information Security Management Act (FISMA)**, **Network & Information Systems (NIS)**) for CI security to protect against both physical and cyber threats.

It is inevitable that regulatory bodies take a more assertive stance. The ever-increasing sophistication of threats, coupled with the interconnectivity of global systems, necessitates a robust regulatory framework that can provide both guidance and enforcement to secure vital assets.

These enhanced regulations and standards will be designed not only to establish a baseline of security practices but also to foster a culture of continuous improvement in the security posture of CIs. They will cover a broad spectrum of requirements, from the implementation of specific technologies and processes to the enforcement of industry-wide protocols that address current and emerging threats.

The evolution of regulatory frameworks will require organizations to be agile and responsive. Compliance will no longer be a box-ticking exercise but a dynamic process that integrates security into the very fabric of organizational operations. To this end, regulations are likely to mandate regular assessments, audits, and the disclosure of security breaches, ensuring that security is both transparent and verifiable.

International cooperation will also play a crucial role, as threats to CI are not bound by national borders. Harmonizing standards across countries and regions will be essential to address the global nature of cyber threats and to facilitate the sharing of best practices. International bodies may set these standards, creating a common language and approach to securing CI.

In the wake of these developments, companies will need to invest in their compliance capabilities. This investment will include the development of internal policies, the training of staff, and the implementation of systems that can adapt to changing regulatory demands. The use of automated tools for compliance management will become more widespread, aiding organizations in navigating the complex landscape of regulations.

Cross-sector collaboration

Different sectors of CI (such as energy, telecommunications, and transportation) will increasingly need to collaborate to address shared risks and vulnerabilities.

In the intricate web of modern CI, the lines between sectors are becoming ever more blurred. The energy sector relies on telecommunications for smart grid management, while transportation systems are interwoven with both to ensure smooth and efficient operations. In this interconnected environment, a vulnerability in one sector can quickly become a systemic threat, highlighting the imperative for cross-sector collaboration.

This collaboration is rooted in the understanding that risks and threats are not confined to single sectors. Cyber attackers often exploit the weakest link in a chain that spans multiple industries. For example, an attack on the healthcare sector could also expand and affect the pharmaceutical industry and health-tech vendors. A coordinated approach to security that leverages collective expertise and resources is not just beneficial but essential.

Cross-sector collaboration will involve a range of activities, from joint TI sharing and coordinated **incident response** (**IR**) to collaborative research and development initiatives. The goal is to create a unified front against threats and to foster resilience that is reinforced by the strengths of each sector.

The complexity of such collaboration cannot be understated. It requires alignment of security practices and harmonization of operational technologies and protocols. It demands robust communication channels and the establishment of trust among sectors that may have traditionally operated in silos.

To facilitate this level of cooperation, frameworks and platforms for information sharing will need to be established. These could take the form of formal alliances, industry working groups, or PPPs. They will need to be underpinned by clear guidelines on data privacy, sharing protocols, and collaborative workflows.

Summary

In this chapter, we examined the complexities of securing CI in a timely manner. As CIs such as power grids and transportation networks become more interconnected with technology, they face greater cyber threats. This section highlights the importance of advanced cybersecurity, including AI and robust encryption, and stresses the essential role of human oversight and training. The growing use of IoT devices and smart technologies adds to the complexity, requiring stronger security protocols.

Emphasis is placed on the need for collaboration, including PPPs and cross-sector cooperation, for intelligence sharing and best practices. The text also points out the necessity for resilience and rapid recovery in infrastructure systems to withstand and recover from attacks. Anticipated stricter regulations are discussed as essential for bolstering infrastructure security against diverse threats. We present a future where safeguarding CI involves a blend of technological innovation, collaborative effort, and continuous adaptation.

Conclusion

In conclusion, *Critical Infrastructure Security: Cybersecurity lessons learned from real-world breaches* is more than just a book; it's a comprehensive resource that empowers you to understand, analyze, and protect the CIs that underpin our society. As you close this book, you carry with you not only a wealth of knowledge but also the responsibility and capability to contribute to the security and resilience of our vital systems.

As you close the final chapter of this book, you find yourself not at the end of a journey but at the precipice of a vast and mysterious digital expanse. The pages you have turned have illuminated the shadowy realms of cyber threats and the intricate dance of securing the infrastructures that pulse beneath our modern world's surface.

In your hands, this book has been more than a guide; it has been a key to unlocking cryptic codes and hidden passages of cybersecurity. The knowledge you've gained whispers of unseen battles and silent warriors in the digital night, safeguarding our most precious assets from ghosts that lurk in the machine.

You are now part of this enigmatic world, a guardian armed with insight and understanding, facing a horizon where the lines between technology and intrigue blur. The future of our CI, a complex web of secrets and strategies, beckons you to continue this quest, to uncover the mysteries, and to protect the unseen heartbeats of our civilization.

And, in the words that echo through the galaxy far, far away, *In a dark place we find ourselves, and a little more knowledge lights our way*. As you conclude your journey through this book, remember – like the timeless struggle between light and dark, the quest for cybersecurity is an ongoing saga.

May this book be the light that guides you through the shadows of the digital world, illuminating your path as you navigate the complexities of protecting our most CIs. May the force of knowledge and vigilance be with you, always.

References

To learn more about the topics covered in this chapter, take a look at the following resources:

- *Garcia, A. B., Babiceanu, R. F., & Seker, R. (2021, April). Artificial intelligence and machine learning approaches for aviation cybersecurity: An overview. In 2021 Integrated Communications Navigation and Surveillance Conference (ICNS) (pp. 1-8). IEEE.*

- *Halbouni, A., Gunawan, T. S., Habaebi, M. H., Halbouni, M., Kartiwi, M., & Ahmad, R. (2022). Machine learning and deep learning approaches for cybersecurity: A review. IEEE Access, 10, 19572-19585.*

- *Xin, Y., Kong, L., Liu, Z., Chen, Y., Li, Y., Zhu, H., ... & Wang, C. (2018). Machine learning and deep learning methods for cybersecurity. IEEE Access, 6, 35365-35381.*

- *Ansari, M. F., Dash, B., Sharma, P., & Yathiraju, N. (2022). The impact and limitations of artificial intelligence in cybersecurity: a literature review. International Journal of Advanced Research in Computer and Communication Engineering.*

- *Patil, P. (2016). Artificial intelligence in cybersecurity. International journal of research in computer applications and robotics, 4(5), 1-5.*

- *Shah, V. (2021). Machine Learning Algorithms for Cybersecurity: Detecting and Preventing Threats. Revista Espanola de Documentacion Científica, 15(4), 42-66.*

- *Chehri, A., Fofana, I., & Yang, X. (2021). Security risk modeling in smart grid critical infrastructures in the era of big data and artificial intelligence. Sustainability, 13(6), 3196.*

- *Sakhnini, J., Karimipour, H., Dehghantanha, A., & Parizi, R. M. (2020). AI and security of critical infrastructure. Handbook of Big Data Privacy, 7-36.*

- *Berghout, T., Benbouzid, M., & Muyeen, S. M. (2022). Machine learning for cybersecurity in smart grids: A comprehensive review-based study on methods, solutions, and prospects. International Journal of Critical Infrastructure Protection, 38, 100547.*

- *Dawson, M., Bacius, R., Gouveia, L. B., & Vassilakos, A. (2021). Understanding the challenge of cybersecurity in critical infrastructure sectors. Land Forces Academy Review, 26(1), 69-75.*

- *Sontan, A. D., & Samuel, S. V. (2024). The intersection of Artificial Intelligence and cybersecurity: Challenges and opportunities. World Journal of Advanced Research and Reviews, 21(2), 1720-1736.*

- *Yu, K., Tan, L., Mumtaz, S., Al-Rubaye, S., Al-Dulaimi, A., Bashir, A. K., & Khan, F. A. (2021). Securing critical infrastructures: deep-learning-based threat detection in IIoT. IEEE Communications Magazine, 59(10), 76-82.*

- *Ren, K., Zheng, T., Qin, Z., & Liu, X. (2020). Adversarial attacks and defenses in deep learning. Engineering, 6(3), 346-360.*

- *Morris, J. X., Lifland, E., Yoo, J. Y., Grigsby, J., Jin, D., & Qi, Y. (2020). Textattack: A framework for adversarial attacks, data augmentation, and adversarial training in NLP. arXiv preprint arXiv:2005.05909.*

- *Mani, N., Moh, M., & Moh, T. S. (2021). Defending deep learning models against adversarial attacks. International Journal of Software Science and Computational Intelligence (IJSSCI), 13(1), 72-89.*

- *Newaz, A. I., Haque, N. I., Sikder, A. K., Rahman, M. A., & Uluagac, A. S. (2020, December). Adversarial attacks to machine learning-based smart healthcare systems. In GLOBECOM 2020-2020 IEEE Global Communications Conference (pp. 1-6). IEEE.*

- *Kazim, E., & Koshiyama, A. S. (2021). A high-level overview of AI ethics. Patterns, 2(9).*

- *Borenstein, J., & Howard, A. (2021). Emerging challenges in AI and the need for AI ethics education. AI and Ethics, 1, 61-65.*

- *Morley, J., Machado, C. C., Burr, C., Cowls, J., Joshi, I., Taddeo, M., & Floridi, L. (2020). The ethics of AI in health care: a mapping review. Social Science & Medicine, 260, 113172.*

- *Munoko, I., Brown-Liburd, H. L., & Vasarhelyi, M. (2020). The ethical implications of using artificial intelligence in auditing. Journal of Business Ethics, 167(2), 209-234.*

- *Aizenberg, E., & Van Den Hoven, J. (2020). Designing for human rights in AI. Big Data & Society, 7(2), 2053951720949566.*

- *Langenberg, B., Pham, H., & Steinwandt, R. (2020). Reducing the cost of implementing the advanced encryption standard as a quantum circuit. IEEE Transactions on Quantum Engineering, 1, 1-12.*

- *Tezcan, C. (2021). Optimization of advanced encryption standard on graphics processing units. IEEE Access, 9, 67315-67326.*

- *Teng, L., Li, H., Yin, S., & Sun, Y. (2020). A Modified Advanced Encryption Standard for Data Security. Int. J. Netw. Secur., 22(1), 112-117.*

- *Anajemba, J. H., Iwendi, C., Mittal, M., & Yue, T. (2020, April). Improved advance encryption standard with a privacy database structure for IoT nodes. In 2020 IEEE 9th international conference on communication systems and network technologies (CSNT) (pp. 201-206). IEEE.*

- *Prabhakaran, V., & Kulandasamy, A. (2021). Hybrid semantic deep learning architecture and optimal advanced encryption standard key management scheme for secure cloud storage and intrusion detection. Neural Computing and Applications, 33(21), 14459-14479.*

- Silverman, J. H. (2020). *An introduction to lattices, lattice reduction, and lattice-based cryptography.* Lect. Notes PCMI Grad. Summer Sch.

- Esgin, M. F. (2020). *Practice-oriented techniques in lattice-based cryptography* (Doctoral dissertation, Monash University).

- Botes, M., & Lenzini, G. (2022, June). When cryptographic ransomware poses cyber threats: Ethical challenges and proposed safeguards for cybersecurity researchers. In *2022 IEEE European Symposium on Security and Privacy Workshops (EuroS&PW)* (pp. 562-568). IEEE.

- Perrier, E. (2021). *Ethical quantum computing: A roadmap.* arXiv preprint arXiv:2102.00759.

- Esquibel, J. M. (2023). *Willingness to Partner in Public-Private Partnership for Cybersecurity of Critical Infrastructure* (Doctoral dissertation, Ph. D. dissertation, Dept. Inf. Sci., Naval Postgraduate School, Monterey, CA, USA).

- Cappelletti, F., & Martino, L. (2021). *Achieving Robust European Cybersecurity through Public-Private Partnerships: Approaches and Developments.* Antonios Nestoras, 58.

- Karabacak, B., Ikitemur, G., & Igonor, A. (2020). *A Mixed Public-Private Partnership Approach for Cyber Resilience of Space Technologies.*

- Paek, S. Y., Nalla, M. K., & Lee, J. (2020). Determinants of police officers' support for the public-private partnerships (PPPs) in policing cyberspace. *Policing: An International Journal, 43(5)*, 877-892.

- Ampratwum, G., Tam, V. W., & Osei-Kyei, R. (2023). Critical analysis of risks factors in using public-private partnership in building critical infrastructure resilience: A systematic review. *Construction Innovation, 23(2)*, 360-382.

- Mottahedi, A., Sereshki, F., Ataei, M., Nouri Qarahasanlou, A., & Barabadi, A. (2021). The resilience of critical infrastructure systems: A systematic literature review. *Energies, 14(6)*, 1571.

- Alkhaleel, B. A., Liao, H., & Sullivan, K. M. (2022). Risk and resilience-based optimal post-disruption restoration for critical infrastructures under uncertainty. *European Journal of Operational Research, 296(1)*, 174-202.

- Almoghathawi, Y., González, A. D., & Barker, K. (2021). Exploring recovery strategies for optimal interdependent infrastructure network resilience. *Networks and Spatial Economics, 21*, 229-260.

- Malatji, M., Marnewick, A. L., & Von Solms, S. (2022). Cybersecurity capabilities for critical infrastructure resilience. *Information & Computer Security, 30(2)*, 255-279.

- Alkhaleel, B. A., Liao, H., & Sullivan, K. M. (2022). Risk and resilience-based optimal post-disruption restoration for critical infrastructures under uncertainty. *European Journal of Operational Research, 296(1)*, 174-202.

- Nasiri, M., Ukko, J., Saunila, M., & Rantala, T. (2020). Managing the digital supply chain: The role of smart technologies. *Technovation, 96*, 102121.

- Palermo, S. A., Maiolo, M., Brusco, A. C., Turco, M., Pirouz, B., Greco, E., ... & Piro, P. (2022). Smart technologies for water resource management: An overview. Sensors, 22(16), 6225.

- Schomakers, E. M., Lidynia, C., & Ziefle, M. (2022). The role of privacy in the acceptance of smart technologies: Applying the privacy calculus to technology acceptance. International Journal of Human-Computer Interaction, 38(13), 1276-1289.

- Koch, T., Möller, D. P., & Deutschmann, A. (2020). Smart Technologies as a Thread for Critical Infrastructures. Smart Technologies: Scope and Applications, 275-289.

- Xu, P., Lee, J., Barth, J. R., & Richey, R. G. (2021). Blockchain as supply chain technology: considering transparency and security. International Journal of Physical Distribution & Logistics Management, 51(3), 305-324.

- Syed, N. F., Shah, S. W., Trujillo-Rasua, R., & Doss, R. (2022). Traceability in supply chains: A Cyber security analysis. Computers & Security, 112, 102536.

- Cheung, K. F., Bell, M. G., & Bhattacharjya, J. (2021). Cybersecurity in logistics and supply chain management: An overview and future research directions. Transportation Research Part E: Logistics and Transportation Review, 146, 102217.

- Lee, Y. J. (2020). Defense ICT supply chain security threat response plan. Convergence Security Journal, 20(4), 125-134.

- Min, S. H., & Son, K. H. (2020). Comparative analysis on ICT supply chain security standards and framework. Journal of the Korea Institute of Information Security & Cryptology, 30(6), 1189-1206.

- PN, S. (2021). The impact of information security initiatives on supply chain robustness and performance: an empirical study. Information & Computer Security, 29(2), 365-391.

- Karie, N. M., Sahri, N. M., & Haskell-Dowland, P. (2020, April). IoT threat detection advances, challenges and future directions. In 2020 workshop on emerging technologies for security in IoT (ETSecIoT) (pp. 22-29). IEEE.

- Sharma, B., Pokharel, P., & Joshi, B. (2020, July). User behavior analytics for anomaly detection using LSTM autoencoder-insider threat detection. In Proceedings of the 11th international conference on advances in information technology (pp. 1-9).

- Tan, L., Yu, K., Ming, F., Cheng, X., & Srivastava, G. (2021). Secure and resilient artificial intelligence of things: a HoneyNet approach for threat detection and situational awareness. IEEE Consumer Electronics Magazine, 11(3), 69-78.

- Rangaraju, S. (2023). Ai sentry: Reinventing cybersecurity through intelligent threat detection. EPH-International Journal of Science And Engineering, 9(3), 30-35.

- Zhou, J., Wu, Z., Xue, Y., Li, M., & Zhou, D. (2022). Network unknown-threat detection based on a generative adversarial network and evolutionary algorithm. International Journal of Intelligent Systems, 37(7), 4307-4328.

- *Nikolov, G., Debatty, T., & Mees, W. (2020). Evaluation of a multi-agent anomaly-based advanced persistent threat detection framework. In The Twelfth International Conference on Evolving Internet (INTERNET 2020).*

- *Benabderrahmane, S., Berrada, G., Cheney, J., & Valtchev, P. (2021). A rule mining-based advanced persistent threats detection system. arXiv preprint arXiv:2105.10053.*

- *Shah, V. (2021). Machine Learning Algorithms for Cybersecurity: Detecting and Preventing Threats. Revista Espanola de Documentacion Cientifica, 15(4), 42-66.*

- *Akcay, S., & Breckon, T. (2022). Towards automatic threat detection: A survey of advances of deep learning within X-ray security imaging. Pattern Recognition, 122, 108245.*

- *Asharf, J., Moustafa, N., Khurshid, H., Debie, E., Haider, W., & Wahab, A. (2020). A review of intrusion detection systems using machine and deep learning in internet of things: Challenges, solutions and future directions. Electronics, 9(7), 1177.*

- *Giedraityte, V. (2022). Interinstitutional and Cross-Sectorial Collaboration to Ensure Security. Europe Alone: Small State Security Without the United States, 325.*

Index

packtpub.com

Subscribe to our online digital library for full access to over 7,000 books and videos, as well as industry leading tools to help you plan your personal development and advance your career. For more information, please visit our website.

Why subscribe?

- Spend less time learning and more time coding with practical eBooks and Videos from over 4,000 industry professionals

- Improve your learning with Skill Plans built especially for you

- Get a free eBook or video every month

- Fully searchable for easy access to vital information

- Copy and paste, print, and bookmark content

Did you know that Packt offers eBook versions of every book published, with PDF and ePub files available? You can upgrade to the eBook version at packtpub.com and as a print book customer, you are entitled to a discount on the eBook copy. Get in touch with us at customercare@packtpub.com for more details.

At www.packtpub.com, you can also read a collection of free technical articles, sign up for a range of free newsletters, and receive exclusive discounts and offers on Packt books and eBooks.

Other Books You May Enjoy

If you enjoyed this book, you may be interested in these other books by Packt:

Implementing Palo Alto Prisma Access

Tom Piens Aka 'Reaper'

ISBN: 9781835081006

- Configure and deploy the service infrastructure and understand its importance

- Investigate the use cases of secure web gateway and how to deploy them

- Gain an understanding of how BGP works inside and outside Prisma Access

- Design and implement data center connections via service connections

- Get to grips with BGP configuration, secure web gateway (explicit proxy), and APIs

- Explore multi tenancy and advanced configuration and how to monitor Prisma Access

- Leverage user identification and integration with Active Directory and AAD via the Cloud Identity Engine

Unveiling the NIST Risk Management Framework (RMF)

Thomas Marsland

ISBN: 9781835089842

- Understand how to tailor the NIST Risk Management Framework to your organization's needs
- Come to grips with security controls and assessment procedures to maintain a robust security posture
- Explore cloud security with real-world examples to enhance detection and response capabilities
- Master compliance requirements and best practices with relevant regulations and industry standards
- Explore risk management strategies to prioritize security investments and resource allocation
- Develop robust incident response plans and analyze security incidents efficiently

Packt is searching for authors like you

If you're interested in becoming an author for Packt, please visit `authors.packtpub.com` and apply today. We have worked with thousands of developers and tech professionals, just like you, to help them share their insight with the global tech community. You can make a general application, apply for a specific hot topic that we are recruiting an author for, or submit your own idea.

Share Your Thoughts

Now you've finished *Critical Infrastructure Security*, we'd love to hear your thoughts! Scan the QR code below to go straight to the Amazon review page for this book and share your feedback or leave a review on the site that you purchased it from.

`https://packt.link/r/183763503X`

Your review is important to us and the tech community and will help us make sure we're delivering excellent quality content.

Download a free PDF copy of this book

Thanks for purchasing this book!

Do you like to read on the go but are unable to carry your print books everywhere?

Is your eBook purchase not compatible with the device of your choice?

Don't worry, now with every Packt book you get a DRM-free PDF version of that book at no cost.

Read anywhere, any place, on any device. Search, copy, and paste code from your favorite technical books directly into your application.

The perks don't stop there, you can get exclusive access to discounts, newsletters, and great free content in your inbox daily

Follow these simple steps to get the benefits:

1. Scan the QR code or visit the link below

https://packt.link/free-ebook/9781837635030

2. Submit your proof of purchase

3. That's it! We'll send your free PDF and other benefits to your email directly

www.ingramcontent.com/pod-product-compliance
Lightning Source LLC
Chambersburg PA
CBHW080633060326
40690CB00021B/4920